Hugh Miller
Stonemason, Geologist, Writer

Michael A. Taylor

Foreword by
Marian Allardyce McKenzie Johnston MA MSc

Preface by
Dr David Alston

First published in 2007 by
NMS Enterprises Limited – Publishing

Reprinted in 2022 by
NMS Enterprises Limited – Publishing
a division of NMS Enterprises Limited
National Museums Scotland
Chambers Street
Edinburgh EH1 1JF

Text © National Museums Scotland 2007, 2022
except for quotations from other works, or as credited.

Images © National Museum Scotland and other individually named sources (see page 11).

'Cromarty Man (*c.*1837)' © and reproduced by kind permission
of Dr James Robertson [first published in Jenni Calder (ed.),
Present Poets: Poems for the Museum of Scotland
(NMS Publishing Limited, Edinburgh, 1998)].

No reproduction permitted without written permission.

ISBN: 978-1-910682-35-7

*No part of this publication may be reproduced, stored in a retrieval system
or transmitted, in any form or by any means, electronic, mechanical, photocopying,
recording or otherwise, without the prior permission of the publisher.*

The right of Michael A. Taylor to be identified
as the author of this book has been asserted
by him in accordance with the
Copyright, Designs and Patents Act 1988.

British Library Cataloguing in Publication Data
A catalogue record of this book
is available from the British Library.

Cover design by Mark Blackadder.
Cover images: Detail from an engraving by Reverend Drummond of Edinburgh
after photograph by James Good Tunny, *c.*1855. Image © National Museums Scotland;
Ammonite from Eathie, NMS.G.1859.33.3957. Image © National Museums Scotland;
Eathie. Image © Dr Lyall Anderson.

Printed and bound in Great Britain by
Bell & Bain Ltd, Glasgow.

This product is made of material from well-managed
forests and other controlled sources.

CONTENTS

MAP of Scotland 4
MAP of Edinburgh and Leith in the 1850s 5
FOREWORD by Marian Allardyce McKenzie Johnston MA MSc 6
ACKNOWLEDGEMENTS by Dr Michael A. Taylor 9
IMAGE CREDITS 11
'Cromarty Man (c.1837)' by Dr James Robertson 12
PREFACE by Dr David Alston 13

CHAPTERS

1. INTRODUCTION: *One of the living forces of Scotland* 17
2. *A wild insubordinate boy* 22
3. *A life of manual labour* 27
4. *The literary lion of Cromarty* 34
5. *A sort of Robinson Crusoe in geology* 40
6. *A long, and, in its earlier stages, anxious courtship* 45
7. *A plain working man, in rather humble circumstances* 53
8. *Among the remains of a different creation* 59
9. *Strife, toil, and comparative obscurity* 68
10. *His business was to fight* 75
11. *The truth I speak, impugn it whoso list* 83
12. *The landscape was one without figures* 91
13. *The quiet enthusiasm of the true fossil-hunter* 97
14. *He clothed the dry bones of science* 103
15. *Exceedingly plausible and consummately dangerous* 113
16. *A gray maud, buckled shepherd-fashion* 121
17. *These are but small achievements* 127
18. *A tenderly affectionate parent* 133
19. *Dearest Lydia, dear children, farewell* 141
20. *Life itself is a school* 149

SOURCES 157
   A note on Hugh Miller Williamson's memoir 162
   Glossary 164
   Places and websites to visit 166
   Further reading 168
   Miller's major books 170
INDEX 171

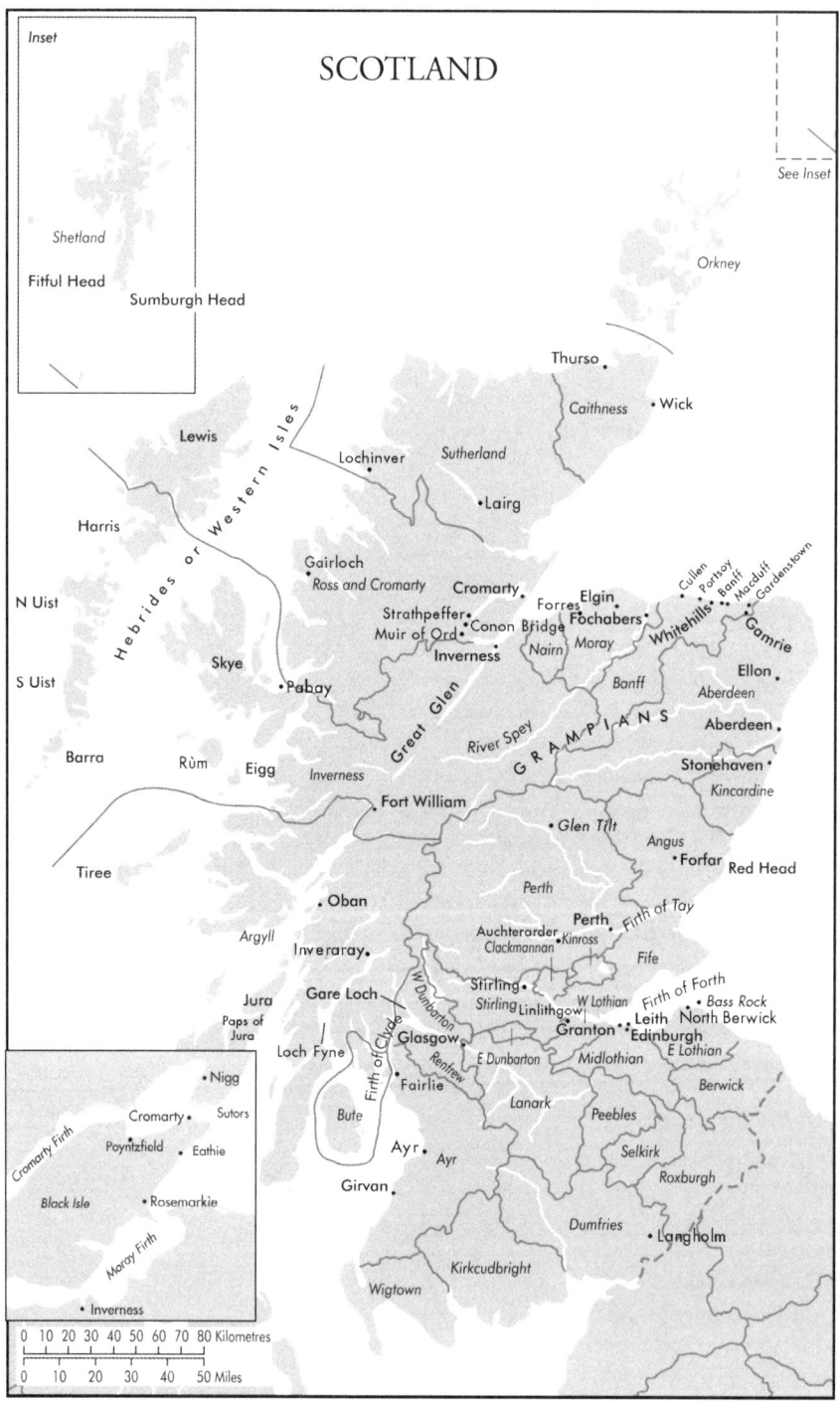

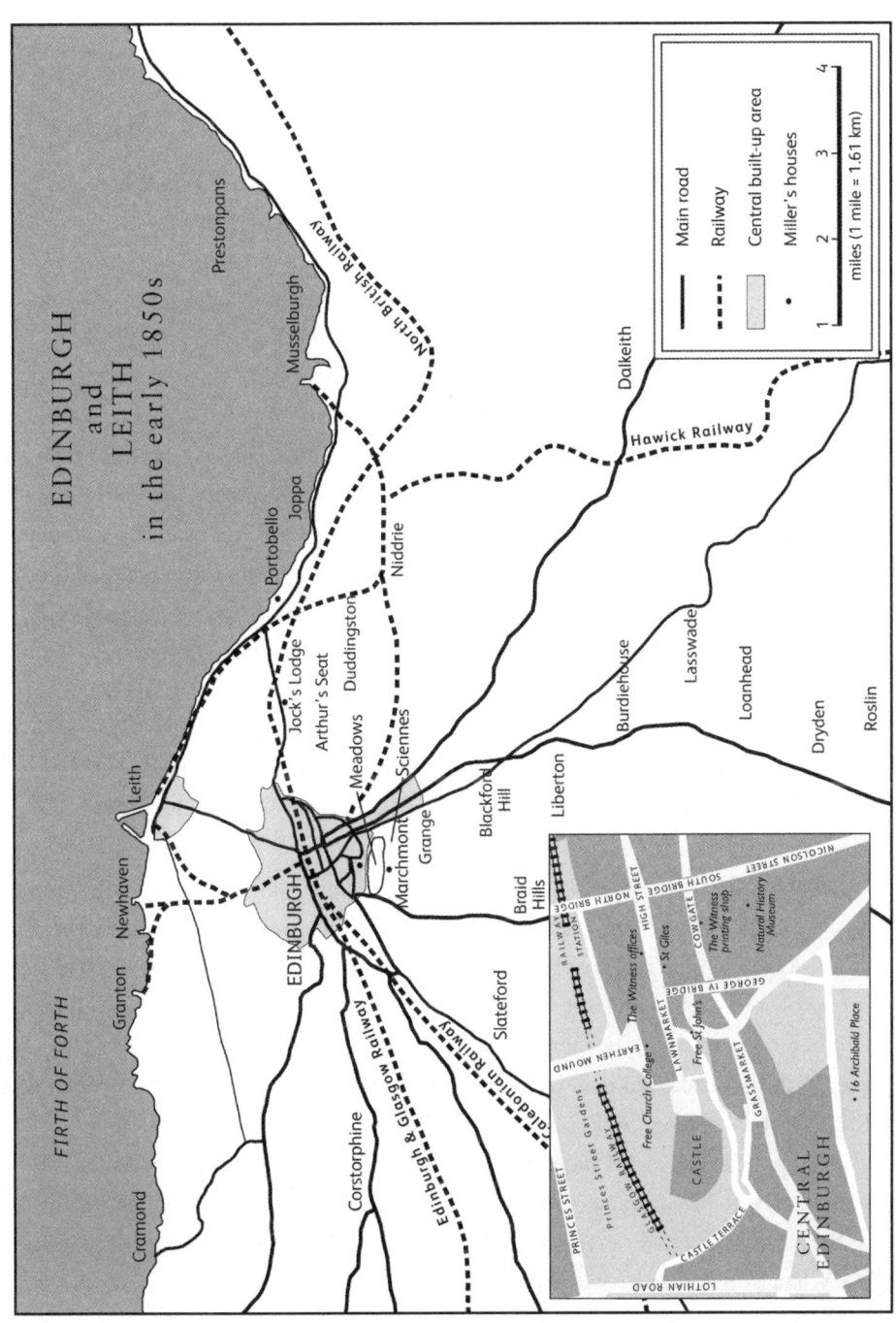

FOREWORD

Marian Allardyce McKenzie Johnston
MA MSc

At the opening of Miller House in Cromarty in April 2004, someone asked me – as a direct descendant of Hugh Miller – whether I felt any connection with him. I was at a loss. Which Hugh? Was he the little lad looking out vainly for the sail of his father's sloop; the voracious reader but unruly schoolboy; the nephew learning from his uncles 'the habit of observation' and listening to their talk of the social, political, and religious interests of the time; the contumacious adolescent who thought he had had enough formal education at 16; the apprentice stonemason who wrote poetry; the fossil collector whose friend, Miss Dunbar, 'was desirous to fix me down to literature as my proper walk; and I, on the other hand, was desirous of escaping into science'? Or was he the lover, who – in order to earn enough to marry – buckled down to learn accountancy? Or the evangelical Presbyterian who felt so strongly about patronage that he became the journalist-editor of an Edinburgh newspaper? So many Hughs.

I think Mike Taylor has achieved the impossible in his new biography. He has brought all the Hughs together to make the complete man. I felt little connection with the Hugh of Peter Bayne's *Life and Letters*, written in the style of his age – Victorian sanctimonious pomposities. Even Hugh's granddaughter wrote 'Hoots-Toots' in the margin of one letter. And I felt no connection whatsoever with the interpretation of Hugh's character by some biographers. But in Mike's Hugh, I recognise Family.

I first heard of Hugh when in 1927 my Middleton Grandfather took me to see the cottage in Cromarty where Hugh was born, told me how he had lost his father at the age of five – my own age at the time – showed me the fossils, and told me what a clever, hard-working man he had been. This Grandfather, Tom Middleton, had married Lydia Davidson, the eldest daughter of Hugh's elder daughter Harriet. The Middletons had come to the Black Isle as farmers at the end of the eighteenth century – recruited by 'George Ross, the Scotch Agent' who wished (as Hugh put it in *Scenes and Legends*) 'to initiate his tenantry in the art of rearing wheat'. They had thrived; most of them married Cromarty girls. So I found that Tom's mother was Esther Ross Taylor, whose father, Walter, had been at the Cromarty School with Hugh Miller. Esther herself had been at school with Harriet Miller. Tom's grandmother was Eliza Allardyce, the daughter of Hugh's friend, Mrs Ann Allardyce, who was descended from George Urquhart of Greenhill. Greenhill was the original name of Rosefarm, where Tom was born, now run by my sister Bright Gordon.

How did Hugh's children – Harriet, William, Bessie and young Hugh – remember him and their home life? Unfortunately little evidence remains. When Hugh's wife Lydia died in 1876 at her daughter Bessie's home, Harriet and her husband, the Revd John Davidson, were in Adelaide, South Australia. William and his wife Maggie were with the army in India. Young Hugh was as yet unmarried. Bessie, married to the Revd Norman Mackay, took all her mother's papers and possessions to their manse near Lochinver and thence to Edinburgh. My grandmother Lydia Middleton asked for, and was sent, a small selection of family letters from the 'trunks and trunks of papers' then still in Bessie's attic. The remainder was later taken by Bessie's eldest son Hugh to his Brighton flat, which received a direct hit from a Luftwaffe bomb during the Second World War and was totally destroyed.

In 1890 Tom Middleton wrote to his future wife Lydia from Baroda College in India, where he had been appointed Professor of Agriculture by the Gaekwar, the ruler of Baroda:

> I was working over one of your Grandfather – Hugh Miller's – books. …
> Your Grandmother must have studied Geology pretty closely for her preface to 'Lectures on Geology'. It shows she was well up in the history and literature of the subject.

This made me look at other prefaces and wonder if Lydia Miller's part in Hugh's posthumous fame had been recognised. I already thought that – after the Cromarty years – she must have been a long-suffering wife. Hugh was a workaholic; his few free hours he spent roaming the hills or the beaches, never 'observing meal times'. He took his holidays by himself; he avoided society unless with other enthusiasts. Lydia as a girl had boarded with George Thomson, 'the friend of Burns', in Edinburgh, and had enjoyed musical soirées and literary conversation. Later she had spent time with her mother's relations at Egham Lodge in Surrey – another social life. This time, life in Edinburgh was different.

After Hugh's death Lydia set about getting his unpublished works into print, despite persistent health problems. Harriet, seventeen when her father died, was 'haunted by his death'. She also had to support her mother and organise the younger children. Her daughter, another Lydia, said she was 'intellectual, not maternal … extraordinarily witty, accomplished, musical'. Harriet, following in her mother's literary footsteps, wrote books for girls, some serialised in *The Adelaide Observer*, for which she also wrote reviews and essays. Her last book, *Sir Gilbert's Children*, written when she was already in her terminal illness, was 'an account of her childhood in Edinburgh'. Her daughter Lydia wrote, 'Papa once told me that she was incapable of writing the plain unadorned truth; she had to embroider everything'. *Sir Gilbert's Children* is heavily embroidered: the Knighthood, elegant clothes, dinner parties, and regular family holidays are perhaps how Harriet would have liked to remember her childhood – but it does show a happy family life, a devoted husband and loving father. Hugh clearly showed great interest in his children's characters and activities – when he was around. Both Harriet and Bessie wrote with great love and admiration of their mother and her 'wonderfully interesting letters' … which went up in smoke. Harriet's daughter Lydia wrote of her grandfather's mother, Harriet Miller, 'she was very occult'; and of Hugh himself, 'of course Grandpapa was a Freak – Pure Genius!' Although some of his descendants have inherited his energies, none has shown his rare combination of so many interests and talents.

To finish, on behalf of those sixty-plus surviving descendants scattered about the world, I should like to express enormous thanks to the anonymous benefactor who has made possible the publication of this well-researched and put-together biography.

(2007)

ACKNOWLEDGEMENTS

Dr Michael A. Taylor
NATIONAL MUSEUMS SCOTLAND

The publication of this book was enabled by a generous donation through the National Museums Scotland Charitable Trust. The Trustees of National Museums Scotland gratefully acknowledge the kind support of the donor, who wishes to remain anonymous.

The book stems in part from research for the exhibition *Testimony of the Rocks: Hugh Miller 1802–1856* put on at the Royal Museum, Edinburgh, from 9 March to 3 June 2002 by the National Museums of Scotland (as it was called then), the National Library of Scotland, and the National Galleries of Scotland. However, this book has benefited enormously from the publications surrounding the 2002 bicentenary, and, as evident from the lists of sources, I am deeply indebted to many colleagues in all disciplines, notably the editors and authors of Shortland (1996) and Borley (2002, 2003), whose new findings and interpretations I have used here and in past publications.

Dr David Alston, Dr Lyall Anderson, Professor Tom Devine, Mr Martin Gostwick, Professor Simon Knell, Mr Henry McKenzie Johnston, Mrs Marian McKenzie Johnston, and Dr Ralph O'Connor kindly reviewed this book in draft, to its considerable benefit. Professor Knell and Dr O'Connor extensively discussed problematical issues and were generous with their insights; and Dr O'Connor kindly provided a MS copy of his forthcoming book. For help with specific points, I am also grateful to Dr Lester Borley, Mr Archie Foley, Professor David Joyce, Professor James Secord, Mr Gavin Sprott, Dr Sara Stevenson, and Mr Jon Watt.

I thank Mrs Marian McKenzie Johnston, Elgin Museum and The Moray Society, the Geological Society of London, the National Library of Scotland, and the National Trust for Scotland, for access and permission to cite manuscript material, Dr James Robertson for permission to reproduce his poem, and Mr

Brian McNeill for permission to reproduce an excerpt from his song, as well as the staffs of the Library of National Museums Scotland, National Library of Scotland, and New College, Edinburgh, for all their support.

I am most grateful to Mrs Marian McKenzie Johnston, a great-great-granddaughter of Hugh Miller, and her husband, Mr Henry McKenzie Johnston, for help, information, encouragement, and critical comment over the years; and to Mrs Frieda Gostwick and Mr Martin Gostwick, successive custodians of Miller's Birthplace Cottage (National Trust for Scotland), and to Dr David Alston, then of Cromarty Courthouse Museum, for their help and advice on my visits to Cromarty, so important for an insight into Miller. Finally, I thank my partner Helen Handoll for her constant support and encouragement of my interest in Miller, started those many years ago when I found my great-uncle's copy of *My Schools and Schoolmasters* in the family bookcase.

Author's note to revised edition of 2022
Some revisions have been made pertaining in particular to Harriet Taylor's memoir; Miller's publisher's name; geological catastrophism; comments on Miller's *Witness* articles at the British Association meeting of 1840; the foundation of the Cottage museum at Cromarty; the source of T. H. Huxley's quotation; and the dates of Bonnar's painting and the publication of *Leading Articles*. Suggestions on places to visit and books to read have also been updated, and three black & white images added within the text. I am grateful to the late Professor J. H. Burns, the late Mr Martin Gostwick, Dr Andrew Ross, Dr Alix Powers-Jones, Dr Janet Trythall, and the late Professor Nigel Trewin for their help, and to the late Mr Henry McKenzie Johnston CB for permission to reproduce copyright material. I thank Ms Lesley Taylor, Mrs Kate Blackadder, Ms Lynne Reilly and Ms Margaret Wilson and their colleagues of NMS Enterprises Ltd – Publishing for their careful and patient endeavours with both editions of this book.

* * *

As this revision went to press, it was suggested in the popular media that Miller held racist and white-supremacist opinions late in life.[1] A robust and detailed rebuttal has been published, confirming and greatly extending the assessment presented here as well as in the original edition.[2]

ACKNOWLEDGEMENTS

NOTES TO ACKNOWLEDGEMENTS

1. The most detailed of these allegations was M. Gostwick 2020. 'Degraded races, hopelessly lost', *Hugh's News. Magazine of the Friends of Hugh Miller* **46**, 11–14.
2. R. O'Connor 2021a, 2021b. 'Hugh Miller: racist or anti-racist? Part 1: slavery, the Clearances and Frederick Douglass', ibid. **48**, www.thefriendsofhughmiller.org.uk/index.asp [unpaginated revised version accesssed 24 January 2022] and 'Part 2: scientific racism and the scale of civilization', ibid. **50**, 2– 23. Especially clear statements of Miller's anti-racist stance occur on pp. 91–96 of this volume, particularly Miller's account of the little Parsi girl. A book-length study by O'Connor is in preparation.

Image credits

All images and photographs are credited individually on the page. Every attempt has been made to contact copyright holders for permission to use material in this book. If any source has been inadvertently overlooked, please contact the publisher.

DR LYALL ANDERSON
art sections 1(6), 2(3), 2(8)

ELGIN MUSEUM
text 82

FREE CHURCH OF SCOTLAND
(Photograph by George Thomson)
art section 2(5)

DR HELEN HANDOLL
art section 1(5)

INVERNESS MUSEUM & ART GALLERY, HIGH LIFE HIGHLAND
text 134; art section 2(8)

MORAY COUNCIL MUSEUMS SERVICE
text 58

NATIONAL GALLERIES OF SCOTLAND
National Galleries of Scotland:
art section 1(3)

Scottish National Portrait Gallery,
Scottish National Photography Collection:
text 28; art sections 1(3), 1(5), 2(1), 2(4), 2(6)

NATIONAL MUSEUMS SCOTLAND
text 16, 52, 74, 120; art sections 1(2x3), 1(6x3), 1(7), 2(1x2), 2(2x5), 2(3x4), 2(5), 2(6), 2(7)

SCRAN (© licensed via SCRAN) for:
Cromarty Courthouse.
Licensor www. SCRAN.ac.uk
art section 1(1x1)

Special Collections, Glasgow University Library. SCRAN
art section 2(4), 2(6), 2(7)

RCAHMS/SCRAN
text 140; art section 1(8)

St Andrews University Library.
Licensor www. SCRAN.ac.uk: art section 1(1x1)

MARIAN SUMEGA
text 5

DR MICHAEL A. TAYLOR
text 102, 150; art sections 1(1), 1(4x4), 1(5), 1(7), 1(8), 2(8)

ADDITIONAL SOURCES

MAP OF SCOTLAND text 4

Adapted from a © map produced by Jim Lewis for Hugh Miller: *The Cruise of the Betsey, with Rambles of a Geologist*, with introduction and notes by Dr Michael A. Taylor (NMS Enterprises Limited – Publishing, 2003, 2021).

With thanks, also, to Cromarty Image Library, and to Dr David Alston, whose assistance in the sourcing of images was greatly appreciated.

Cromarty Man
(*c*.1837)

Hugh Miller splits another prehistoric page:
Bird tracks like pressed flowers, fish bones
That are diagrams of death, in a book made long
Before his God first thought of him. Miller
Recreates the species, draws lines between them:
No humanoid amphibians crawl from slime,
No blue-arsed Lord Monboddo swings through trees,
Pursued by midwives. But the old coastline beyond
The receded sea rises like the half-dry knowledge
Of his half-drowned fear; and his poised hammer
Is the relic of that pirate ancestor whose ghost
He saw in childhood once, above him on the stair.

James Robertson, 1998

PREFACE

Dr David Alston
Councillor, Black Isle Ward
HIGHLAND COUNCIL

It is a pleasure to welcome the publication of this biography of Hugh Miller, both in my formal capacity as the local Highland councillor and, more personally, as a resident of the town of Cromarty – the place which shaped Miller's early life and which has, ever since, been influenced by his writings and reputation. Dr Taylor's thoroughly researched book is a timely reassessment of what we know of Miller the man, and of what can be said of his place both in Scottish history and in the history of science. The balanced appraisal presented in *Hugh Miller: Stonemason, Geologist, Writer* will, I am sure, be well received, not least in his home town.

Hugh Miller and Cromarty have had an enduring, but complex, relationship. When Miller published his autobiographical *My Schools and Schoolmasters*, first as a series of articles in the *Witness* and then, in 1854, in book form, he presented the detail of life in early nineteenth-century Cromarty to readers throughout Scotland. Consequently, the people of the ancient burgh lived, for a time, with a public profile probably greater than any other small town in the country. This was, however, an image of Cromarty's prosperous, but now vanished, past. Mid-nineteenth-century Cromarty was a changed place and, on his visits home, Miller found the town becoming a 'second deserted village' – an allusion to Goldsmith's poem 'The Deserted Village'. Others painted an even bleaker picture. As early as the year of the Disruption, 1843, one visitor to the town, Alexander Oswald Brodie, who was employed at the building of Cromarty's lighthouse, wrote in his diary that it was 'a miserable place in the last stages of decay'. And by the time of Miller's death in 1856, two years after the publication of *My Schools and Schoolmasters*, the community had fallen on even harder times.

Cromarty was, of course, still immensely proud of Miller and enthusiastic about the erection of a memorial to its most famous son. But his reputation, which did not diminish with his death, was also a badly needed asset for the decaying town – a reason why, with its traditional sources of employment in drastic decline, Cromarty might be promoted as 'a place of summer resort'. There was, however, not only the problem of marrying the expectations of prospective visitors – the readers of *My Schools and Schoolmasters* – with the reality, but the more general difficulty of rising expectations. One author had already described the 'tender skinned, tender legged, tender stomached' tourists along the Caledonian Canal, who 'demanded well-aired, flea-less sheets, chicken broth and no trouble'. Thus Cromarty, as a tourist destination, struggled with its ruinous houses, middens, pigs, and the pervasive smell of fish. Nevertheless, correspondents in the local newspapers floated their ideas of how the town should promote itself, perhaps the most extravagant being the notion of a 'hydropathic establishment' where 'sick people might wander with Hugh Miller's work in hand over the classic ground', which might be considered 'the Italy of Scotland' lying as it did between two expanses of water, with 'Cromarty like Brindisi in the heel of the boot'. Even if such grand schemes were not realised, a number of visitors did come, and some, at least, found that 'wherever one turns the spirit of Hugh Miller is to be found, fresh and fragrant as ever'.[1]

Cromarty has continued to live with, and make use of, its association with Miller. Early postcards of the town, produced in greater numbers once the firth became a naval base in 1913, often featured (all on the same card) a local view, a broad tartan border, a sprig of white heather, and the image of Hugh Miller's craggy face – used (as one might today put it) to brand Cromarty, just as Robert Burns became Ayr's unique selling point. Cromarty streets had already been renamed after Miller (Miller Road and Miller Lane), and both the town's first *Visitors' Guide* (1890) and the first album of local photographs (1900) featured his birthplace and monument. Over a century later, the cottage where he was born still attracts tourists; and all those who enter Cromarty today are greeted by road signs proclaiming it as the 'Birthplace of Hugh Miller, geologist and writer'.

Sometimes the Hugh Miller brand – unfranchised and so available to all – has had downright surprising uses. The local team, formed in the 1880s to play football to the new Association rules, was the Hugh Miller Football Club. And when the Black Isle Brewery, in association with the National Trust for Scotland, produced a Hugh Miller Ale in 2002, to mark the bicentenary of his birth, Miller became probably the first person in history to be honoured by both a beer and a temperance organisation (Cromarty's late nineteenth-century Hugh Miller Lodge of the Good Templars).

There were, of course, from time to time, serious attempts to reassess Miller's work, such as *Hugh Miller: A Critical Study*, published in 1905 by William Mackay Mackenzie, another eminent son of Cromarty who was well placed to do the job. But, nevertheless, the 'real Hugh Miller' has, at times, been in danger of disappearing, whether obscured by an image designed for tourists, lost to indifference, or distorted by fashionable misinterpretations of his life. For example, the portrayal of Miller as a man caught tragically between his faith and the emerging results of his geological studies – between the biblical account of creation and the scientific concept of biological evolution – has persisted, and was the central theme of Reginald Barrett-Ayres' 1970s opera. It has – as Dr Taylor shows – no foundation in fact.

Cromarty has not been unconscious of these ironies, ambiguities and misinterpretations – and chose, as the logo for the bicentenary celebrations in 2002, the image of a fossil and the words 'Hugh Who?' Dr Taylor has, through his own writings, and the assistance he has provided others, played an important role in the general revival of interest in Miller – and he gives us now an authoritative and accessible answer to that question.

(2007)

NOTES TO PREEFACE

1. Cromarty quotations can be found in Alston, D. *My little town of Cromarty: the history of a northern Scottish town* (Edinburgh, 2006), pp 264, 266 and 268.

HUGH MILLER

Frontispiece in W. M. Mackenzie (ed.), *Selections from the Writings of Hugh Miller* (Paisley, 1908), engraved by unknown artist after photograph by A. Rae and Sons, Banff. (For Rae photograph, see e.g. 10401518 on http://www.scienceandsociety.co.uk, downloaded 6 November 2006.)

(IMAGE © NATIONAL MUSEUMS SCOTLAND)

CHAPTER 1

INTRODUCTION
One of the living forces of Scotland[1]

IN 1902 the elderly geologist Archibald Geikie drew a vivid word-picture from his distant youth:

> Among the picturesque figures that walked the streets of Edinburgh in the middle of last century, one that often caught the notice of the passer-by was that of a man of good height and broad shoulders, clad in a suit of rough tweed, with a shepherd's plaid across his chest and a stout stick in his hand. His shock of sandy-coloured hair escaped from under a soft felt-hat; his blue eyes, either fixed on the ground or gazing dreamily ahead, seemed to take no heed of their surroundings. His rugged features wore an expression of earnest gravity, softening sometimes into a smile and often suffused with a look of wistful sadness, while the firmly compressed lips betokened strength and determination of character. The springy elastic step with which he moved swiftly along the crowded pavement was that of the mountaineer rather than of the native of a populous city. A stranger would pause to look after him and to wonder what manner of man this could be. If such a visitor ventured to question one of the passing townsmen, he would be told promptly and with no little pride 'That is Hugh Miller'. No further description or explanation would be deemed necessary, for the name had not only grown to be a household word in Edinburgh and over the whole of Scotland, but had now become familiar wherever the English language was spoken, even to the furthest western wilds of Canada and the United States.[2]

Here, it seems, was captured the essence of a man who symbolised Scotland to his contemporaries even in his visage and walk. But here also arise questions about the subject and the author. In the 1850s Hugh Miller had encouraged the young Geikie's first steps in his scientific career. But now, in 1902, Geikie had already retired from the Directorship-General of the Geological Survey, and would soon ascend another peak of British science, the Presidency of the Royal Society of London. He was, moreover, the author of books, on the history of geology, which are today perceived as classic examples of the author unashamedly portraying the past as a heroic era, and adding lustre to the science of his own day by representing it as the product of a glorious progress. Nor was Geikie the only one to crown Miller with laurels. In *Self-help* (1859), the book which today symbolises Victorian values, Samuel Smiles portrayed Miller as a secular saint of self-education, perseverance and hard work. And in 1902, Revd Principal Robert Rainy of the Free Church College asserted that 'Hugh Miller had enriched, dignified, and ennobled … Scotland as a whole, and to his memory was due whatever could be done worthily to express the veneration and regard with which they cherished his name'.[3]

In his afterlife Miller was evidently ripe for adoption as a great Scottish hero, used by those who followed, to validate their own perceptions of themselves, their times, and their origins. Yet who was Hugh Miller that he should be so glorified, even sainted, from the moment of his tragically early death in 1856? How did he present himself to his contemporary public – as any writer must, especially one whose writings had a strong autobiographical streak? And what are we to make of him in today's Scotland, so notoriously a land of 'no gods and precious few heroes'? This quotation from Hamish Henderson's poem sequence, 'Elegies for the dead in Cyrenaika', is the title of both Christopher Harvie's history of modern Scotland and Brian McNeill's folksong attacking the abuse of Scottish history. The song's chorus runs:

> 'Cause there's no gods and there's precious few heroes
> But there's plenty on the dole in the land o' the leal
> And it's time now to sweep the future clear
> Of the lies of a past that we know was never real.[4]

Of course, that Hugh Miller was a Victorian hero doesn't necessarily mean that what we know of his life was a lie, or that he was a spuriously romantic hero of the likes of 'Bonnie Prince Charlie' – for whose royal Stuart dynasty Miller had, in any case, the greatest contempt. But one must still be critical, even sceptical, when assessing the Miller whom historians and writers have presented to us. This is partly because they have struggled to make sense of what can seem a mass of contradictions. Miller can certainly appear an odd hero. A stonemason who became a famous geologist; a school dropout who came to edit one of Scotland's leading newspapers; a layman who helped found the Free Church in the fight for Scottish religious liberty; a progressive who opposed universal suffrage; a devout Presbyterian Christian who insisted that scientific evidence was as validly God-given as the Bible; a man who attacked both religious obscurantism and evolutionary theory; an expounder of self-help to whom the environmentalist John Muir and the capitalist robber baron Andrew Carnegie both paid tribute; a scientist who created some of Scotland's greatest writing, yet produced barely any scientific papers; a Lowlander who spoke out for the Highlanders against the lairds; and, above all, a man who was in many respects selfless and without personal social ambition, and yet earned the lasting respect of the Scottish people – Hugh Miller was all of these.

The resolution of this problem lies in projecting ourselves back into his world, just as Miller took his readers deep into the teeming coal forests:

> … long withdrawing lakes, fringed with dense thickets of the green Calamite, tall and straight as the masts of pinnaces, and inhabited by enormous fishes, that glittered through the transparent depths in their enamelled armour of proof; or glades of thickest verdure, where the tree-fern mingled its branch-like fronds with the hirsute arms of the gigantic club-moss, and where, amid strange forms of shrub and tree no longer known on earth, the stately Araucarian reared its proud head two hundred feet over the soil … .

> The great size and marvellous abundance of those … fishes of the Carboniferous period may well excite wonder. … If the gar-pike, a fish from three to four feet in length, can make itself so formidable … that even the cattle and horses that come to drink at the water's side are scarce safe from its attacks, what must have been the character of a fish … from thirty to forty feet in length, furnished with teeth thrice larger than those of the hugest alligator … ?[5]

Like those denizens of the Carboniferous, Miller must be seen in his own world to be understood. His Scotland, about which he wrote copiously and cogently, is identifiably continuous with modern Scotland, and yet sometimes seems utterly alien: not so much that it believed in heroes and in God, but also in its complete freedom for the monied, combined with astonishing exploitation of workers at labour and at home, in a country that was still riven linguistically and culturally.

In their depth and range, Miller's achievements justify our paying critical attention to his story. In this book I attempt a fresh synoptic look at Miller, a 'natural history' of the man in the context of his time. I seek to tell a coherent story, as Miller himself tried to do with his own life in *My Schools and Schoolmasters*, and as his widow sought to do, through Peter Bayne, in *The Life and Letters of Hugh Miller*. For simplicity, I present Miller's life as a river braided into distinct but intermeshing streams of work, family life, religion, science, and so forth, although he would of course have perceived it as a single flow. This is not a novel, and I remain silent where there are gaps in the historical evidence. But we are fortunate to have Miller's own character, thoughts and actions preserved in his writings, of which I use substantial excerpts, 'word-pictures' that reveal the man himself, with the grit and ink of his labours under his fingernails, just as he himself brought the past to life for his contemporaries. (I have edited his quotations only minimally, to avoid destroying his characteristic rhythm; unusual words, other than minor variant spellings, can be found in the glossary or the endnotes.) From all this I seek to reveal the unjustly neglected figure of Hugh Miller, not as some Jekyll-and-Hyde-like curiosity, but as the man whom the Victorians hugely respected for his consistency, integrity and independence.

Dr Michael A. Taylor
EDINBURGH, 2007

NOTES TO CHAPTER 1

1. A. Geikie, *Scottish reminiscences* (Glasgow, 1904), p. 377.
2. Anon. *The centenary of Hugh Miller being an account of the celebration held at Cromarty on 22nd August, 1902* (Glasgow, 1902). pp. 40–41.
3. Anon., *Centenary*, p. 13.
4. C. Harvie, *No Gods and Precious Few Heroes* (Edinburgh, 1998), p. [x].
5. H. Miller, *Sketch-book of popular geology* (Edinburgh, 1889) pp. 85, 184.

NOTES TO WHOLE BOOK

Measures and prices

Modern units are used throughout this book; however, measurements taken from Miller's writings are left in their original units to avoid spurious accuracy. One foot of twelve inches is about 0.30 metres; one yard of three feet about 0.91 metres; one fathom of six feet about 1.83 metres; and one (land) mile of 1760 yards about 1.61 kilometres. One gill is 0.14 litres (on Imperial standard; old Scots measures varied). One hundredweight of 112 pounds is about 51 kilogrammes.

Until 1971 the pound sterling comprised 20 shillings (abbreviated to 's'), each of 12 (old) pence ('d'); thus £1 = 20 s and 1s = 12 d. The pound was worth far more in Miller's time than today, almost unmeasurably so given the different standards and ways of living and the relative prices of items. However, a multiplier of 150–200 x gives some idea of value. £3.50 in 2007 would then equate to fourpence halfpenny in 1840 – which was the price of the *Witness* (and other newspapers of that time). This amount was easily afforded by a middle-class professional with an annual income of, perhaps, several hundred pounds, but not by the many families in Victorian Britain which subsisted on less than a pound a week.

Maps

A map of Scotland is provided on page 4. On this, the old county names are shown in italics, but towns and other locations in roman. Thus the county of Inverness or Inverness-shire is marked as '*Inverness*', but the town is 'Inverness'. A map of Edinburgh district is also provided on page 5. A range of (older) maps of Scotland can be found on www.nls.uk.

Church denominations

The Presbyterian Churches of Scotland, including the Reform and Secession Churches, have undergone various mergers and splits since 1843. Therefore, at least in the sense of institutional continuity, the division between the Church of Scotland (which has long abandoned patronage) and the several Free Churches no longer corresponds to that of the Disruption.

CHAPTER 2

A wild insubordinate boy[1]

DURING the eighteenth century the small burgh of Cromarty was one of northern Scotland's most important trading and industrial towns. It lay at the end of the Black Isle, which is in fact a peninsula rather than an island, and on the superb natural harbour of the Cromarty Firth. Geographically in the Gaelic-speaking Highlands, Cromarty was culturally and linguistically in the Lowlands, the northernmost outpost of Scots (and English) speaking Lowlanders. And here, on 10 October 1802, in a thatched cottage in Cromarty, a son was born, Hugh, to Harriet and Hugh Miller. The young Hugh had Gaelic-speaking relatives and some Gaelic forebears, but was a Lowlander Scot by upbringing and by mentality. As an adult his reading aloud from a book was 'remarkably pure and distinct, little tainted with any provincial accent. Except for the peculiar pronunciation of some of the vowels, it would have been nearly free from this,' remembered his widow.[2] But maybe Miller was modulating his speech to her preferences, for other, perhaps more objective, listeners recalled the Black Isle inflections in his lectures: 'The axe had been busy in the glades' became 'The exe hed been bussy in the gleds', and 'the bitter cup of affliction' 'the butter kip of affluction'.[3]

Miller's father, also called Hugh, was a coastal shipmaster, sufficiently successful to build a new stone house, today called Miller House, next to the cottage. He died in 1807 when his ship was lost. His son would

> … climb, day after day, a grassy protuberance of the old coast-line immediately behind my mother's house, that commands a wide reach of the Moray Firth, and to look wistfully out, long after every one else had ceased to hope, for the sloop with the two stripes of white and the two square topsails.[4]

The family stayed in the cottage, dependent on the widow's work as a seamstress, rent from letting the new house, and help from relatives. But Miller was not left fatherless. His maternal uncles James, a harness-maker, and Alexander ('Sandy'), a cartwright, undertook 'the work of instruction and discipline' so that he 'owed to them much more of my real education than to any of the teachers whose schools I afterwards attended'.[5] Miller's interests took after both. James loved local legends and traditions, while Sandy was especially interested in nature. Miller went walking with Sandy along the beach, collecting interesting pebbles and crystals, and exploring shore life. He also enjoyed wandering the countryside around Cromarty, sometimes meeting Sandy at his sawpit in the woods.

A bright child, Miller soon learned to read for himself. As he grew, his reading widened to include travels such as Anson's *Voyages*, old Scots patriotic epics such as Blind Harry's *Wallace* and Barbour's *Bruce*, poetry such as Pope's *Odyssey*, and novels such as *Gulliver's Travels*. Even as a child he became so keen on telling stories to his friends – a skill perhaps learnt partly from his mother and other relatives – that he was nicknamed 'Sennachie' (Gaelic for a clan bard, but here meaning simply a storyteller).

As Miller grew older he did not do as well at school as his uncles hoped or his intelligence allowed, partly because the parish schoolmaster could not control his charges. Miller became something of a juvenile *litterateur* dabbling in poetry (and later, when he was 18, composing a handwritten 'newspaper', the *Village Observer*, which only ever had one copy). Nonetheless, he found it difficult to apply himself to school work, such as Latin, which he found uninteresting. By his middle teens, he was somewhat unruly, roaming around with other youths and getting into minor scrapes. His uncles were greatly frustrated, while

> ... my poor mother, despairing of a reformation in my conduct gave me up altogether, and suffered her hopes and her affections to rest on my two sisters, who were interesting little girls, and both docile and intelligent. Alas! in the winter of 1816, when the one was in her tenth and the other in her twelfth year, they were both seized at nearly the same time by a malignant fever, then raging in the place, and died within a few days of each other. The stroke made a considerable impression on me, for I was not destitute of affection. I remember being wrung to the heart by overhearing my mother

remark how different her condition would have been had it pleased Heaven to have taken her son from her, and left one of her daughters.[6]

But any effect on his behaviour did not last; perhaps he resented his mother's marriage in 1819 to Andrew Williamson, a nail-maker.

> I had ... become a wild insubordinate boy, and the only school in which I could properly be taught was that world-wide school which awaited me, in which Toil and Hardship are the severe but noble teachers.[7]

His uncles still felt he was fitted for an academic training. Despairing of the parish school, they tried a private school in Cromarty, but Miller still got into trouble, as when in self-defence he stabbed a schoolfellow in the thigh. Finally, one day, confused by his own Scottish pronunciation and by the differing practice of his previous teachers, Miller found himself confronting the dominie:

> ... when required one evening to spell the word '*awful*', with much deliberation – for I had to translate, as I went on, the letters *a-w* and *u* – I spelt it word for word, without break or pause, as a-w-f-u-l. 'No', said the master. 'a-w, *aw*, f-u-l, *awful*; spell again.' This seemed preposterous spelling. It was sticking in an *a*, as I thought, into the middle of the word, where, I was sure, no *a* had a right to be; and so I spelt it as at first. The master recompensed my supposed contumacy with a sharp cut athwart the ears with his tawse; and again demanding the spelling of the word, I yet again spelt it as at first. But on receiving a second cut, I refused to spell it any more; and, determined on overcoming my obstinacy, he laid hold of me and attempted throwing me down. As wrestling, however, had been one of our favourite ... exercises, and as few lads of my inches wrestled better than I, the master, though a tall and tolerably robust fellow, found the feat considerably more difficult than he could have supposed. We swayed from side to side of the school-room, now backwards, now forwards, and for a full minute it seemed to be rather a moot point on which side the victory was to incline. At length, however, I was tripped over a form; and as the master had to deal with me, not as master usually deals with pupil, but as one combatant deals with another, whom he has to beat into submission, I was mauled in a way that

filled me with aches and bruises for a full month thereafter. I greatly fear that, had I met the fellow on a lonely road five years subsequent to our encounter ... he would have caught as sound a thrashing as he ever gave to little boy or girl in his life; but all I could do at this time was to take down my cap from off the pin, when the affair had ended, and march straight out of school. And thus terminated my school education.[8]

Thus Miller left school, aged around 16.[9] In so doing he lost the real opportunities open in Scotland to intelligent boys from even relatively poor families to go to university and train for a profession. His mother's remarriage now prompted him to take up a trade; no doubt he did not want to live with Williamson any more than Williamson wanted to support an idle stepson. As it so happened, Miller had noticed how one of his Gaelic cousins, who lived at Lairg, and who worked as a stonemason and slater, managed to indulge his wider literary interests in the winter close season. So Miller 'determined on being a mason. I remembered my Cousin George's long winter holidays, and how delightfully he employed them; and, by making choice of Cousin George's profession, I trusted to find, like him, large compensation, in the amusements of one-half the year, for the toils of the other half'.[10] Miller hoped – he later said – that in so doing he could make a career in literature and, perhaps, natural science. But maybe Miller was rationalising the mess he had made of his life.

Both my uncles, especially James, were sorely vexed by my determination to be a mason; they had expected to see me rising in some one of the learned professions; yet here was I going to be a mere operative mechanic, like one of themselves! I spent with them a serious hour, in which they urged that, instead of entering as a mason's apprentice, I should devote myself anew to my education. Though the labour of their hands formed their only wealth, they would assist me, they said, in getting through college; nay, if I preferred it, I might meanwhile come and live with them: all they asked of me in return was that I should give myself as sedulously to my lessons as, in the event of my becoming a mason, I would have to give myself to my trade. I demurred. The lads of my acquaintance, who were preparing for college had an eye, I said, to some profession; they were qualifying themselves to be lawyers, or medical men, or, in much larger part, were studying for the

Church; whereas I had no wish, and no peculiar fitness to be either lawyer or doctor; and as for the Church, that was too serious a direction to look in for one's bread, unless one could honestly regard one's-self as *called* to the Church's proper work; and I could not. There, said my uncles, you are perfectly right: better to be a poor mason – better to be anything honest, however humble – than an *uncalled* minister. ...

Slowly and unwillingly my uncles at length consented that I should make trial of a life of manual labour.[11]

NOTES TO CHAPTER 2

1. H. Miller, *My schools and schoolmasters* (Edinburgh, 1905), p. 141.
2. L. M. F. F. Miller, 'Mrs Hugh Miller's Journal', *Chambers's Journal* (6)**5**, p. 371.
3. Alston, 'The fallen meteor', p. 227; W. M. Mackenzie, 'Biographical introduction', pp. v–xii in *Schools*, p. xii.
4. *Schools*, p. 26.
5. *Schools*, p. 35.
6. H. Miller (ed. M. Shortland), *Hugh Miller's Memoir* (Edinburgh, 1995) pp. 104–5.
7. *Schools*, p. 141.
8. *Schools*, pp. 143–44.
9. Miller is unclear when this combat occurred: maybe late spring 1819, as it was before his mother remarried (5 June) and 'some months' before starting work (February 1820), unless he downplayed his idleness (*Memoir*, esp. pp. 105–8; *Schools*, pp. 129–51).
10. *Schools*, p. 151.
11. *Schools*, pp. 152–53.

CHAPTER 3

A life of manual labour[1]

IN February 1820, Miller started to learn the trade of a country stonemason as an apprentice of his mother's brother-in-law, David Williamson. Initially the gang worked in and around Cromarty, quarrying the stone and working it in the do-everything manner of the country mason. He found the work heavy until he grew into it, but enjoyed the countryside.

Williamson, unable to obtain new contracts, was compelled to work as a journeyman and, around June 1821, Miller went with him to experience the life of a peripatetic mason away from home, first working near Muir of Ord, and then on Conan Mains farm steading south of Conon Bridge. The 1822 season was spent first at Conan Mains, and then in the unpleasant work of rebuilding a ha-ha in a wet ditch at Poyntzfield. Here Miller first found himself suffering chest pains and coughing up bloody mucus, symptoms of the characteristic stonemason's lung disease caused by stone dust. When the job ended in November, he had completed his apprenticeship, and so recuperated at home over the winter and spring, spending part of his time building a cottage for his Aunt Jenny.

Miller developed real pride in his strength and skill as a stonemason. Masons were élite craftsmen who made Scotland's fine buildings, architect-designed churches and grand houses. But they also made the ordinary buildings of burgh and farm which are, in their way, just as notable for the workmanship and sense of proportion which went into their building. A mason had a range of chisels to work stone, driving them with a mell or maul, a large mushroom-headed mallet, whose wooden head absorbed some of the shock. He laid out and checked his work with callipers, rules, set squares and plumb bobs. The mason cut and dressed every block by hand. In cheap work he cut the stones roughly, and filled the gaps with mortar. But in the best work, he made blocks of stone that fitted together precisely. All work was done by hand without power tools, except

HUGH MILLER, 1802–56. GEOLOGIST AND AUTHOR [detail]
David Octavius Hill and Robert Adamson

A calotype photograph of Miller by the pioneering Edinburgh photographers, Hill and Adamson. (Calotypy was the first negative/positive process of photography, using chemically sensitised paper and a negative.) This calotype was taken in 1844, long after Miller last wrought as a mason; but he and his opinions, were marked for life by his time in the lower strata of society.

(SCOTTISH NATIONAL PORTRAIT GALLERY)

perhaps a manual derrick for lifting heavy blocks. Large blocks, even of granite, were often split by wedges driven into a series of holes made with nothing more than an iron bar called a jumper, thumped down and twisted round with sand in the bottom of the hole to wear away the stone. It is indeed our loss that Miller never wrote at length about his craft.

For the 1823 season Miller again wrought at Conan Mains. He then went to Gairloch on the west coast, to work at the manse and later at the nearby Flowerdale Inn. At Gairloch the 'minister himself, a stout, portly, good looking man',[2] instructed the gang to fit up their own accommodation in what had been a hay barn and was now

> ... merely a roof-covered tank of green stagnant water, about three-quarters of a foot in depth, which had oozed through the walls from an over-gorged pond in the adjacent court, that in a tract of recent rains had overflowed its banks, and not yet subsided. Our new house did look exceedingly like a beaver-dam, with this disadvantageous difference, that no expedient of diving could bring us to better chambers on the other side of the wall. My comrade, setting himself to sound the abyss with his stick, sung out in sailor style, 'three feet water in the hold'. Click-Clack broke into a rage: 'That a dwelling for human creatures!' he said. 'If I was to put my horse intil't, poor beast! the very hoofs would rot off him in less than a week. Are we eels or puddocks, that we are sent to live in a loch?' Marking, however, a narrow portion of the ridge which dammed up the waters of the neighbouring pool whence our domicile derived its supply, I set myself to cut it across, and had soon the satisfaction of seeing the general surface lowered fully a foot, and the floor of our future dwelling laid bare. Click-Clack, gathering courage as he saw the waters ebbing away, seized a shovel, and soon showed us the value of his many years' practice in the labours of the stable; and then, despatching him for a few cart-loads of a dry shell-sand from the shore, which I had marked by the way as suitable for mixing with our lime, we had soon for our tank of green water a fine white floor. 'Man wants but little here below', especially in a mason's barrack.[3]

But when the hay harvest came in, the minister ejected them from the barn and they had to move to the cow-house. Nevertheless, that Gairloch puddock-pond was not much worse than the typical tumbledown shed where masons on a job lived for months on end:

> These barracks or bothies are almost always of the most miserable description. I have lived in hovels that were invariably flooded in wet weather by the overflowings of neighbouring swamps, and through whose roofs I could tell the hour at night, by marking from my bed the stars that were passing over the openings along the ridge: I have resided in other dwellings of rather higher pretensions, in which I have been awakened during every heavier night-shower by the rain-drops splashing upon my face where I lay a-bed. I remember that Uncle James, in urging me not to become a mason, told me that a neighbouring laird, when asked why he left a crazy old building standing behind a group of neat modern offices, informed the querist that it was not altogether through bad taste the hovel was spared, but from the circumstance that he found it of great convenience every time his speculations brought a *drove of pigs* or a *squad of masons* the way. And my after experience showed me that the story might not be in the least apocryphal, and that masons had reasons at times for not touching their hats to gentlemen.[4]

Bothy life was crudely masculine in its physical humour and, not least, in its housekeeping standards. The grossly unbalanced diet largely comprised oatmeal, served as cakes roasted on fireside stones, porridge boiled in a pot, and simple brose perhaps made with cold water alone:

> … the food is of the plainest and coarsest description: oatmeal forms its staple, with milk, when milk can be had, which is not always; and as the men have to cook by turns, with only half an hour or so given them in which to light a fire, and prepare the meal for a dozen or twenty associates, the cooking is invariably an exceedingly rough and simple affair. I have known mason-parties engaged in the central Highlands in building bridges, not unfrequently reduced, by a tract of wet weather, that soaked their only fuel the turf, and rendered it incombustible, to the extremity of eating their oatmeal raw, and merely moistened by a little water, scooped by the hand from a neighbouring brook. I have oftener than once seen our own

supply of salt fail us; and after relief had been afforded by a Highland smuggler – for there was much smuggling in salt in those days, ere the repeal of the duties – I have heard a complaint from a young fellow regarding the hardness of our fare, at once checked by a comrade's asking him whether he was not an ungrateful dog to grumble in that way, seeing that, after living on fresh poultices for a week, we had actually that morning got porridge with salt in it.[5]

Miller greatly valued the winter break at home, for he found bothy life depressingly uncivilised, and prone to drink and dissipation:

A life of toil has, however, its peculiar temptations. When overwrought, and in my depressed moods, I learned to regard the ardent spirits of the dram-shop as high luxuries: they gave lightness and energy to both body and mind, and substituted for a state of dulness and gloom, one of exhilaration and enjoyment. Usquebaugh was simply happiness doled out by the glass, and sold by the gill. The drinking usages of the profession in which I laboured were at this time many: when a foundation was laid, the workmen were treated to drink; they were treated to drink when the walls were levelled for laying the joists; they were treated to drink when the building was finished; they were treated to drink when an apprentice joined the squad; treated to drink when his 'apron was washed'; treated to drink when 'his time was out'; and occasionally they learned to treat one another to drink.[6]

During one such celebration in his first year as an apprentice, Miller was so shocked by the effect of a whole gill of whisky, especially on his ability to read, that thereafter he stayed temperate – although, as he admitted, the cost was also a factor.[7]

Work was, at this time, scarce in the north but abundant in Edinburgh and the south. Moreover, Miller was now of age to sell an inherited slum property in Leith, so he took ship there in May 1824. He soon found work at Niddrie House near Edinburgh. Some of his local colleagues were unwelcoming to Miller, seeing him as a hardworking Highlander there to steal their work. They resented his unwillingness to join in their drinking habits, often leaving him to work single-handed at even the hardest tasks. No doubt they reacted against what they saw as

Miller's priggish attitude, but he was genuinely horrified at how his fellow workmen sometimes dissipated their wages in drunken bouts after the fortnightly pay day.

Miller lodged with a local couple. He enjoyed evening walks, seeing his first hedgehogs, and fossil-hunting in nearby quarries. But he thought the agricultural and mining workers of the district ignorant, irreligious and neglected by Kirk and laird, and

> ... the enclosed state of the district, and the fence of a rigorously-administered trespass-law, serious drawbacks ... it is rather to be regretted than wondered at that there should be often less true patriotism in a country of just institutions and equal laws, whose soil has been so exclusively appropriated as to leave only the dusty high-roads to its people, than in wild open countries. ...[8]

Miller was fascinated by Edinburgh's spectacular division into Old and New Towns, and attended the sermons of notable ministers, but found the theatre disappointing compared with his own reading and visualisation of the plays. He witnessed the great fire of Parliament Close, and lingered unsuccessfully in Castle Street to catch a glimpse of Walter Scott.

That winter, employers in the district reduced stonemasons' wages more than would be expected by the reduced working hours of the shorter days. Miller was cajoled into attending a meeting to discuss strike action, but was unimpressed by the first half of the meeting and the way in which his colleagues then repaired to a badger-baiting dram shop and missed the rest. But Miller was biased against trade unions, partly by personal prejudice as he admitted, despite the personal example and exhortation of an old friend from Nigg, William Ross, who was active in the housepainters' union. Miller claimed that such combination (as then called) led to the bullying of other workmen, just as he himself had been in a petty way. He was, in any case, pessimistic about the prospects for success, and doubtful of union organisers' motives and competence. But such views were commonly held about the then unproven labour movement.

Miller earned good money at Edinburgh while the local labour market was at a peak, but had a relapse of his lung illness, the 'stone-cutter's malady' – which, his fellow masons assured him, was his fault for not drinking nearly enough whisky. Wisely, he stopped work and took ship for Cromarty in March 1825.[9]

Miller had been unlucky in embarking on the stonemason's trade just when the Cromarty economy slumped, and then in having his lungs so quickly damaged for life. But his experiences marked him in other ways which would prove useful. He remained proud of the credentials as a skilled working man which he would later deploy in his writings. He gained an understanding, rare within the middle classes, of the precariousness of working-class life, and a sense of outrage at how the wealthy treated the people who lived and worked on their land, although he never overcame his suspicion of the new labour movements. Miller had brought the sensitivities of a Cromarty burgher to 'the tenebrious halo'[10] of proletarian ignorance, intemperance and poverty which surrounded the polite glories of post-Enlightenment Edinburgh. His outlook was fundamentally changed.

NOTES TO CHAPTER 3

1. *Schools*, p. 153.
2. *Memoir*, p. 170.
3. *Schools*, pp. 260–61; Click-Clack was the gang's carter and the ironical quotation is from Goldsmith's poem 'The Hermit'.
4. *Schools*, p. 192.
5. *Schools*, pp. 192–93; the 'turf' is peat.
6. *Schools*, p. 158; 'usquebaugh' was whisky. Presumably apron-washing drinking celebrated completion of the apprentice's first substantial spell of work – maybe even a whole season; his 'time' would be 'out' when his apprenticeship ended.
7. *Memoir*, p. 115.
8. *Schools*, pp. 311–12.
9. He was on 'the smack *Lizzard*', 28 March 1825; Bayne, P. *The life and letters of Hugh Miller* (London, 1871), vol. 1, p. 156; NLS MS.7522.
10. *Schools*, p. 319

CHAPTER 4

The literary lion of Cromarty[1]

AFTER a spell of rest and recovery, no doubt under his mother's care, Miller set up in Cromarty as a monumental mason, carving fine work such as gravestones. This was more skilled but lighter work than that of an ordinary mason, and much less dusty than working en masse with other masons. Work was initially slow, and Miller had time to carve a sundial for his uncles, which today can be found in the garden of his birthplace. But trade did come. Gravestones by Miller, bearing his plain serif lettering often coupled with a scalloped edge to the slab, survive in Cromarty, and are also to be seen in the nearby kirkyards of Nigg and Rosemarkie.

By undertaking such jobs, often for a few days away from home, Miller got to know more of the countryside, its people and their traditions. He also enjoyed the change from the stonemason's dusty workshed to the peace of a burying-ground. Miller humorously likened himself to the hero of Scott's novel *Old Mortality* who travelled the country to keep fresh the memory of the Covenanting martyrs of Stuart persecution by carving their eroded gravestones afresh. And while Miller incised *his* words, people would come and keep him company, from 'a poor idiot boy',[2] to Cromarty's new minister, the Revd Alexander Stewart, who would stay for hours on end. Clearly, Miller was no ordinary mason.

Miller used his spare time to develop his literary skills: reading, practising his writing in his characteristically minute script – perhaps to save paper and postage – and copying his correspondence into a 900-page letter book. However, in the summer of 1828 he could not obtain enough work as a mason and so he moved for a brief while to Inverness where he tried to exploit his literary skills to develop a reputation for good taste, which in turn – he hoped – would promote his trade as a monumental mason. He failed to get his poems into the *Inverness Courier*, a local newspaper, so he committed some of his savings to publishing a volume of his poetry in 1829. *Poems, Written in the Leisure Hours of a Journeyman Mason* was an expensive flop, however, losing around £30 – enough to keep Miller in intellectual austerity for a year or two.³

Miller reached another dead end when he sought patrons to say a good word for his writings, notably Principal Baird of the University of Edinburgh. Baird suggested that Miller try his hand in the Edinburgh literary world, and told him to send him a short account of his life. Miller was reluctant to commit himself to Edinburgh, but optimistically sent Baird an overly long autobiography (published many years later as *Hugh Miller's Memoir*) in 1829 and 1830, together with a crate of his unsold volume of poems. Miller had mishandled a difficult situation.

Miller's poetical fiasco convinced him, correctly, that his skills and his future lay in prose. Moreover, it introduced Miller to Robert Carruthers, the editor of the *Inverness Courier*, who gave him a modest but successful start in journalism as the paper's part-time correspondent back in Cromarty. From 1829 into the 1830s, Miller wrote on a wide range of events, topics and controversies. Amongst the more notable pieces was a series on the herring fishery, which sold well when reprinted as a booklet, *Letters on the Herring Fishery in the Moray Frith*, in 1829. Miller was particularly delighted to learn that Walter Scott had sought a copy. Here is Miller waiting in an open boat at night, a drift-net set from a row of buoys:

> ... A singular appearance attracted my notice. 'How,' said I to one of the boatmen ... 'how do you account for that calm silvery spot on the water, which moves at such a rate in the line of our drift?' He started up. A moment after he called on the others to rise, and then replied: 'That moving speck of calm water covers a shoal of herrings. If it advances a hundred yards farther in that direction, we shall have some employment for you.' ... I hung over the gunwale watching the nets as they approached the side of the boat. The three first, from the phosphoric light of the water, appeared as if bursting into flames of a pale green colour. The fourth was still brighter, and glittered through the waves while it was yet several fathoms away, reminding me of an intensely bright sheet of the aurora borealis. As it approached the side, the pale green of the phosphoric matter appeared as if mingled with large flakes of snow. It contained a body of fish. 'A white horse! a white horse!' exclaimed one of the men at the cork baulk; 'lend us a haul.' I immediately sprung aft,

laid hold on the rope, and commenced hauling. In somewhat less than half an hour we had all the nets on board, and rather more than twelve barrels of herrings. ...

Soon after sunrise the mist began to dissipate, and the surface of the water to appear for miles around roughened as if by a smart breeze, though there was not the slightest breath of wind at the time. ...

The whole frith at this time, so far as the eye could reach, appeared crowded with herrings; and its surface was so broken by them as to remind one of the pool of a waterfall. They leaped by millions a few inches into the air, and sunk with a hollow plumping noise, somewhat resembling the dull rippling sound of a sudden breeze; while to the eye there was a continual twinkling, which, while it mocked every effort that attempted to examine in detail, showed to the less curious glance like a blue robe sprinkled with silver.[4]

Some ships leaving Cromarty carried emigrants to the New World, often Highlanders newly 'cleared' by the lairds, or jumping before they were pushed. Miller's description of one such departure in 1831 was chosen to be carved by the sculptor Richard Kindersley on the Emigration Stone, erected on Cromarty Links in 2002: 'The *Cleopatra*, as she swept past the town of Cromarty, was greeted with three cheers by crowds of the inhabitants, who lined the shore, and the emigrants returned the salute, but, mingled with the dash of the waves and the murmurs of the breeze, their faint huzzas seemed rather sounds of wailing and lamentation, than of a congratulatory farewell'.[5]

Miller's writing was not as developed and assured as it would later become. Many of his letters are perhaps too obviously literary practice pieces to suit that modern taste which prefers the more informal and relaxed style which he later developed. However, his writing was already a tribute to this school dropout's self-education. His characteristic prose style was rooted in the clear, sonorous and classical Augustan diction of the eighteenth-century writers, such as Addison, Pope, Swift, and Goldsmith. But, at least going by his allusions and quotations, he was also much influenced by the Bible in its Authorised Version, and by the Calvinist divines and their historians. One must also add Walter Scott, not for his High Tory politics, but for his historical novels with their fascination with past landscape and history as a way to explain, and place in its context, a rapidly changing Scotland – although Miller would go far beyond Scott by exploring the lost worlds of a pre-human past.

Miller became established as a minor local celebrity for his interest in local history, and also his wider literary activities. One visitor investigating his own family history, John Barkley, consulted Miller in August 1833, and reported back to his family:

> There is living in Cromarty a common stone mason named Hugh Miller, a self educated young man, who has published a volume of poetry, a history of the herring fishery, and also a history of the antiquities of Cromarty. I had a long interview with him and found him a most sagacious, intelligent and upright man. ... This Hugh Miller visits everyone in Cromarty, and the neighbourhood, and is quite the *lion* of that part.[6]

Plainly there was more to Miller's local reputation than the mere curiosity value of any literary workman, real as this last may have been, especially to outsiders – as is hinted in his self-deprecatingly ironical account of how, one day in summer 1831, this 'literary lion'[7] attracted the interest of one Lydia Falconer Fraser. Miss Fraser, newly come to Cromarty to join her widowed mother after staying with relatives in England,

> ... was, I saw, very pretty; and though in her nineteenth year at the time, her light and somewhat *petite* figure, and the waxen clearness of her complexion, which resembled rather that of a fair child than of a grown woman, made her look from three to four years younger. And as if in some degree still a child, her two lady friends seemed to regard her. She stayed with them scarce a minute ere she tripped off again; nor did I observe that she favoured me with a single glance. But what else could be expected by an ungainly, dust-besprinkled mechanic in his shirt sleeves, and with a leathern apron before him? Nor *did* the mechanic expect aught else; and when informed long after, by one whose testimony was conclusive on the point, that he had been pointed out to the young lady by some such distinguished name as 'the Cromarty Poet', and that she had come up to her friends somewhat in a flurry, simply that she might have a nearer look of him, he received the intel-

ligence somewhat with surprise. All the first interviews in all the novels I ever read are of a more romantic and less homely cast than the special interview just related; but I know not a more curious one.[8]

On further acquaintance Miller found that, as well as the country walks he also enjoyed, Miss Fraser had a liking for the 'severer walks of literature'.[9] She had read much less than him, so he simply became a convenient 'sort of dictionary of fact, ready of access, and with explanatory notes attached, that became long or short just as she pleased'.[10]

Miller himself was settling into his role as a Cromarty townsman and burgher, well enough respected locally. He was not cash-rich, but his ownership of his birthplace cottage, and the house which his father had built, gave him the property qualification to become a voter under the 1832 Reform Bill. He was even persuaded to run, successfully, as a burgh councillor, but was so unimpressed by the lack of real business at his first meeting that thereafter he – irresponsibly, one might feel – 'stayed away from the Council board, and did nothing whatever in its behalf, with astonishing perseverance and success, for three years together'.[11]

Here one comes to the crucial question: what did Miller see as his place and role in society? Scottish society, certainly in a small town like Cromarty, had an egalitarian streak. However, in Cromarty as elsewhere, income and custom were increasingly differentiating the classes. Miller's position was ambiguous from the start. He had been born into the middle class (or what at any rate would become it), but his father's death, and his own boyhood diversions, had sent him into what might now be called the upper working classes. Of course Miller understood, and would insist, that to be a working man was not necessarily to be intellectually, morally or spiritually inferior. And in some ways, especially for someone as shy and diffident as Miller, his life was easier and freer, and less bound by social custom, for not taking on the full panoply of a middle-class role (which he could not afford, anyway). Yet this unusual mason was a welcome guest at the tea-tables of the more intellectually minded members of Cromarty's independent middle class of traders, merchants, and people living quietly on their pensions or private means. Perhaps he recognised that, in this sense, he had not dropped in social position by becoming a mason. He was certainly no more attracted to the working-class friendly societies of Cromarty than to the company of working men in Edinburgh. However, this surely reflects more his solitary nature. He retained his ability to mix with the working classes, with fishermen and craftsmen as much as ministers, merchants and bluestockings – and with those literary correspondents outwith Cromarty who genuinely shared his interests and treated him as an equal, such as Sir Thomas Dick Lauder of Edinburgh and Morayshire and Miss Dunbar of Forres. In his writing, too, Miller was establishing what would become a lifelong role as a commentator. He was in the scene yet separate from it, an observer perhaps even more than an actor.

NOTES TO CHAPTER 4

1. *Life*, vol. 1, p. 243.
2. *Life*, vol. 1, p. 373.
3. *Memoir*, pp. 203–4, 212; Miller did benefit from living at home, and possibly from any rents from Miller House which didn't go to his mother.
4. 'Letters on the herring fishery', pp. 148–200 in *Tales and Sketches* (Edinburgh, 1889), pp. 180–85.
5. 'Emigration', p. 44 in Miller, H. (ed. Martin Godstewick) *A noble smuggler and other short stories* (Inverness, 1997).
6. B. Hill, *The remarkable world of Frances Barkley: 1769 1845* (Sidney, British Columbia, 1978), pp. 186–87. Dr David Alston kindly gave this reference and discussed Miller's local reputation, particularly the point that the title of 'The Cromarty Stonemason' is not known to have been used whilst he was working as a mason.
7. *Life*, vol. 1, p. 243.
8. *Schools*, pp. 498–99.
9. *Schools*, p. 501.
10. *Schools*, p. 501.
11. *Schools*, p. 496.

CHAPTER 5

A sort of Robinson Crusoe in geology[1]

BY 1830 Miller was seriously interested in the rocks, and especially their contained fossils, around Cromarty. As often happens with geologists, his progress to such an engagement had been slow, and probably intermittent. As a child he apparently took no particular interest in fossils. It is perhaps only with hindsight that one can see anything special in his childhood collecting of pebbles and minerals on the beach, or in a first uncomprehending encounter with local fossils when he collected the oil-rich Jurassic shale for campfires, thinking naïvely that it was sea coal. Nor, contrary to what is sometimes thought, did Miller's trade as a mason automatically lead him to the delights of palaeontology. Fossils usually make for poor building stone and Miller was not especially likely to come across fossils in the stone which he quarried and worked, beyond the occasional stray find.

Miller did, to be sure, notice, and wonder at, phenomena such as water ripple-marks preserved in sandstone:

> The entire surface was ridged and furrowed like a bank of sand that had been left by the tide an hour before. I could trace every bend and curvature, every cross hollow and counter ridge, of the corresponding phenomena: for the resemblance was [no] half resemblance – it was the thing itself ... . But what had become of the waves that had thus fretted the solid rock, or of what element had they been composed? I felt as completely at fault as Robinson Crusoe did on his discovering the print of the man's foot on the sand.[2]

This suggests an inquisitive mind ripe for what seems his true palaeontological epiphany, one day early in 1820 when Miller was still 17, and the crew was quarrying near the coast at Navity, some three kilometres south of Cromarty.

That day Miller casually broke open one of the pieces of grey nodular limestone washed up on the nearby shore (and not in the quarry they were working). He was amazed to find an ammonite:

> Wonderful to relate, it contained inside a beautifully finished piece of sculpture, – one of the volutes, apparently, of an Ionic capital; and not the far-famed walnut of the fairy tale, had I broken the shell and found the little dog lying within, could have surprised me more. Was there another such curiosity in the whole world? I broke open a few other nodules of similar appearance, – for they lay pretty thickly on the shore, – and found that there might. In one of these there were what seemed to be the scales of fishes, and the impressions of a few minute bivalves, prettily striated; in the centre of another there was actually a piece of decayed wood. Of all Nature's riddles, these seemed to me to be at once the most interesting and the most difficult to expound. I treasured them carefully up, and was told by one of the workmen to whom I showed them, that there was a part of the shore about two miles farther to the west where curiously-shaped stones, somewhat like the heads of boarding-pikes, were occasionally picked up. ....[3]

When work was over for the day, Miller followed his colleague's directions to the 'boarding-pikes' – in fact, belemnites. Here, at Eathie, he first started to read the book in which the past was written:

> The layers into which the beds readily separate are hardly an eighth part of an inch in thickness, and yet on every layer there are the impressions of thousands and tens of thousands of the various fossils peculiar to the Lias. We may turn over these wonderful leaves one after one, like the leaves of a herbarium, and find the pictorial records of a former creation in every page: scallops, and gryphites, and ammonites, of almost every variety peculiar to the formation, and at least some eight or ten varieties of belemnite; twigs of wood, leaves of plants, cones of an extinct species of pine, bits of charcoal, and the scales of fishes; and, as if to render their pictorial appearance more striking, though the leaves of this interesting volume are of a deep black, most of the impressions are of a chalky whiteness. I was lost in admiration and astonishment .....[4]

To Miller, fossils and human artefacts, geology and archaeology, and folklore and tradition, all taken together, told one continuous story of the past. He was interested in them all, but he would come to like fossils best.

Miller's skills as a mason would help him rip up and lay open rocks with ease. But it is unclear how quickly Miller's fossilising actually developed, for his later, and retrospective, writing tends to merge years and telescope time in its narrative flow. Clearly, however, Miller's engagement with the science was – or at least would become – twofold. First, geology was something to do, observe and think about, by exploring the rocky landscape and collecting fossils. But second, geology was something to write about: raw material for his literary activities. These two modes of Miller's engagement could, and apparently did, develop at different paces, and that makes it still harder to judge their growth. We know, for instance, that Miller was, at least occasionally, looking for fossils while at Niddrie around 1824–25, and, when back in Cromarty, he discussed his Eathie finds with his boyhood friend John Swanson. But pursuing his new hobby out in the field didn't necessarily mean that Miller was writing about it – or, if he was writing about it, he wasn't publishing the results. Significantly, he omitted geology from an 1828 list of intended literary and artistic projects, and from his autobiographical *Memoir* of 1829–30. Possibly he did not feel confident that he could write convincingly about geology, or that it was a fit subject for a literary set-piece. Barkley's letter cited above suggests that Miller's local reputation, right into the early 1830s at least, was for literature and not geology, which for now was apparently a private hobby.

Unable to buy expensive books, Miller apparently got no coherent view of the geology of his time. This is hardly surprising, given the vagaries of what ended up in his hands in Cromarty, sometimes decades after it was written. But, in any case, especially by the time it got to Miller, this earliest nineteenth-century geology was a mixture of philosophy, religion, speculation and empirical observation. Swanson did try to pass on any relevant ideas which he encountered at King's College, Aberdeen where he was studying for the ministry – but, alas, natural science was not taught there, and Miller gained little from such things as this notion of a separate creation of pre-adamite humans:

> There was one special hypothesis which he had heard broached, and the utter improbability of which I was not yet geologist enough to detect, which

for a time filled my whole imagination. It had been said, he told me, that the ancient world, in which my fossils, animal and vegetable, had flourished and decayed – a world greatly older than that before the Flood – had been tenanted by rational, responsible beings, for whom, as for the race to which we ourselves belong, a resurrection and a day of final judgment had awaited. But many thousands of years had elapsed since that day – emphatically the *last* to the Pre-Adamite race – had come and gone.[5]

Another bit of philosophical flotsam, previously washed up at Cromarty, was *Telliamed*. This mid-eighteenth-century book outlined a theory of the Earth's formation that included an evolutionary view of the origin and diversification of life. Its speculations, however extreme, did at least sit alongside some sensible observations. No doubt both, in their ways, encouraged Miller in his developing habit of observing the rocks and thinking for himself – a trait he would show all his life.

It was, in any case, one day in 1830 that Miller made a most serendipitous discovery. If unexpected, it did at least stem from what one might nowadays call a research programme prompted by his finds at Eathie. Eathie is south of Cromarty, around the corner, formed by the South Sutor headland where the Cromarty Firth opens into the sea. Miller reasoned that the 'granitic gneiss' of the South Sutor appeared to have been thrust up through the overlying strata. Therefore the same strata as on the Eathie side of the Sutor might occur on the Cromarty side – and be (one imagines) far more convenient for collecting than the long trek to Eathie and the fossil-laden return up a steep hill. So he started searching. Soon, in the little bay between Cromarty and the Sutor, he spotted strata promisingly like those which yielded fossils at Eathie:

> I set myself carefully to examine. The first nodule I laid open contained a bituminous-looking mass, in which I could trace a few pointed bones and a few minute scales. The next abounded in rhomboidal and finely-enamelled scales, of much larger size and more distinct character. I wrought on with the eagerness of a discoverer entering for the first time in a *terra incognita* of wonders. Almost every fragment of clay, every splinter of sandstone, every limestone nodule, contained its organism, – scales, spines, plates, bones, entire fish; but not one organism of the Lias could I find, – no ammonites,

no belemnites, no gryphites, no shells of any kind: the vegetable impressions were entirely different; and not a single scale, plate or ichthyodorulite could I identify with those of the newer formation. I had got into a different world, and among the remains of a different creation … . I wrought on till the advancing tide came splashing over the nodules, and a powerful August sun had risen towards the middle sky; and, were I to sum up all my happier hours, the hour would not be forgotten in which I sat down on a rounded boulder of granite by the edge of the sea, when the last bed was covered, and spread out on the beach before me the spoils of the morning.[6]

It would be years before Miller, that 'Robinson Crusoe in geology, cut off for years from all intercourse with his kind',[7] fully realised what he had found.

NOTES TO CHAPTER 5

1. Miller, H. *The Old Red Sandstone, or, New walks in an old field* (7th ed., Edinburgh, 1899 printing), p. 139.
2. *Old Red Sandstone*, pp. 38–39.
3. *Old Red Sandstone*, pp. 40–41.
4. *Old Red Sandstone*, pp. 41–42.
5. *Schools*, p. 369.
6. *Old Red Sandstone*, pp. 130–32.
7. *Old Red Sandstone*, p. 139.

CHAPTER 6

A long, and, in its earlier stages, anxious courtship[1]

LYDIA Fraser's mother was the widow of a failed Inverness merchant. She and Lydia had come to Cromarty to live quietly on a modest income eked out by fees from the small private school for girls which Lydia operated. Mrs Fraser became concerned as Hugh Miller and Lydia grew closer, for she considered him 'no very fitting mate'[2] for her middle-class daughter – a view doubtless informed by her own family's financial travails. Accordingly, she forbade their meeting, but Lydia omitted to tell Hugh, and so their woodland trysts continued.

Eventually, however, in 1833 Mrs Fraser agreed to a three-year engagement while Miller sought better employment in Scotland, and if he was unsuccessful they might then emigrate to America with a gift of £100 from her. By late 1834 Miller was beginning to take the prospect of emigration seriously, although, as Lydia recalled many years later, he was reluctant to leave his native land:

> His temperament was the least sanguine I ever knew. ... He loved to feel himself every inch a Scotsman. It was from Scottish history he drew his earliest inspiration. The scenery, the traditions, the very soil of his country were inexpressibly dear to him. They were an immense part of himself, of his inner life; and his affections, his predilections, his prejudices even, were strong, constant, unalterably firm. To root them up and transplant them elsewhere was a kind of painful death.[3]

Miller felt that his only realistic alternative to emigration was to become a newspaper editor (this being a time when newspapers had little more than editors and printers in the way of permanent staff). However, he fretted about having to promulgate a fixed political position regardless of his own views, especially as he was not happy with any of the established political parties (this led him some years later to reject an offer to edit an Inverness newspaper). In 1834 Miller published a booklet, *The Traditional History of Cromarty*; and he now decided to attempt a full-length treatment to demonstrate his literary abilities. But one day Miller was asked to call on Robert Ross, a successful Cromarty merchant and ship-owner, and agent-to-be for a new branch of the Commercial Bank, Cromarty's first bank:

> … I was not a little surprised, after we had taken a quiet cup of tea together, and beaten over half-a-dozen several subjects, to be offered by him the accountantship of the branch bank. After a pause of a full half-minute, I said that the walk was one in which I had no experience whatever – that even the little knowledge of figures which I had acquired at school had been suffered to fade and get dim in my mind from want of practice – and that I feared I would make but a very indifferent accountant. I shall undertake for you, said Mr. Ross, and do my best to assist you. All you have to do at present is just to signify your acceptance of the offer made. I referred to the young man who, I understood, had been already nominated accountant. Mr. Ross stated that, being wholly a stranger to him, and as the office was one of great trust, he had, as the responsible party, sought the security of a guarantee, which the gentleman who had recommended the young man declined to give; and so his recommendation had fallen to the ground. 'But *I* can give you no guarantee,' I said. 'From you,' rejoined Mr. Ross, 'none shall ever be asked.' And such was one of the more special *Providences* of my life; for why should I give it a humbler name?[4]

In November 1834 Miller travelled by way of Edinburgh for several months' training at the bank's Linlithgow branch before starting at Cromarty. He found bank work something of a mental strain, but soon realised that it

> … furnished me with an entirely new and curious field of observation, and formed a very admirable school. For the cultivation of a shrewd common sense, a bank office is one of perhaps the best schools in the world. Mere cleverness serves often only to befool its possessor. He gets entangled among his own ingenuities, and is caught as in a net. But ingenuities, plausibilities,

special pleadings, all that make the stump-orator great, must be brushed aside by the banker. The question with him comes always to be a sternly naked one: – Is, or is not, Mr ——— a person fit to be trusted with the bank's money?[5]

Miller was amazed how accurately Ross could assess his customers – a skill which he emulated at least sufficiently to avoid any misplaced loans when Ross was away. Miller was fascinated by economics in action, especially as Cromarty was beginning the long decline that has fortuitously preserved its townscape today much as he knew it. Some shops thrived while others, apparently just as good, failed; honesty and judgement paid off in the end; a farm on poor land invariably failed to thrive; and it was increasingly difficult to make a good business of a small farm.

Many years later, Harriet Taylor, one of Ross's daughters and a pupil of Lydia's, charmingly recalled those days when Miller usually had tea with them each evening. Her sister Isabella always went to fetch him, reappearing mounted on his shoulder:

The meal [over,] he and my father seated themselves on either side of the fire, Mary sat on her father's knee and Isa on Mr Miller's[,] and I sat between them. Mr Miller was the principal speaker but my father spoke too – he was a highly intelligent man and well read, especially in history ancient and modern; had a most retentive memory, and having been in the navy in early life had seen a good deal of the world. We were much interested in Mr Miller's account of all he had seen in Linlithgow; for indeed he had eyes to see both men and nature, and had observed more in one year [sic], and in one small town, than many would have done in a journey round the world. Before leaving school on Friday Miss Fraser [i.e. Lydia] gave a subject on which each girl had to write, the paper to be given in on Monday morning; and I had a standing invitation to go to the office to show Mr Miller mine that he might comment on it before I made a clear copy. I well remember how I went in very quietly, and sat silent by the fireside till his long column of figures was summed up, when he would turn round, and taking my paper, would read it with interest; and carefully point out to me where I might have done better. In the same way when I had finished a drawing I was never satisfied until he had examined it; I wonder at his kindness now. Mr Miller had a very correct eye, and gave a neatness and finish to anything he himself drew. When occasionally my father went from home Mr Miller slept in our house to safeguard the Bank money; and when he did so Betty allowed me to remain out of bed and to sit at table with him while he partook of supper. On one of those evenings a large packet was brought to him, and opening it he said, 'Those are the proofs of my "Scenes and Legends"'; and picking out some leaves he handed them to me saying, 'My dear, read that story; you may yet like to say that yours were the first eyes that saw my book in print'.[6]

This was Miller's first full-length book, *Scenes and Legends of the North of Scotland; or The Traditional History of Cromarty* (1835). It is a remarkable work. On one level, Miller simply drew upon his native hinterland, and indeed his own home, for raw material for a literary work, relying particularly on his mother and other relatives:

> I found it a new and untrodden field, full of those interesting vestiges of past times which are to be found – not in the broken remains of palaces and temples, but in the traditional recollections of the common people.[7]

But on another level, he sought to preserve the history encapsulated within that world: from traditional histories to standing stones and other now silent relics, and the shadowy realm of legend, superstition and the supernatural. There is, for example, a splendidly arbitrary tale of a meteor which each night fell as if to crash into a firthside cottage, and was nevertheless warded off by the cry of a cockerel, until a watcher on a nearby ship bought the bird out of curiosity, with the predictable result. Moreover, perhaps more importantly for his future literary career, Miller's history dipped into a still deeper past, for it was here that he first wrote publicly, if briefly, about his fossil-collecting.[8] Here Miller first seamlessly combined the natural fossils of the rock with the artificial fossils of the landscape, alongside the mental fossils of a human community. In so doing he gave Cromarty a boundless cultural richness, as if a collection of jewels set in the district's verdant landscape.

Scenes and Legends is an early example of the collection of folklore. Miller saved these tales from almost certain loss by using them as vehicles for his own burgeoning literary talents of visualisation and sympathy. Did Miller himself, at any rate beyond childhood, believe these tales, as has sometimes been suggested? He would, no doubt, have regarded those tales as real in the sense of being genuine folk relics, and he did retell them with empathy and effect – but he would not have personally believed the meanings attached to those fossils of the mind, and there is no evidence that Miller's own strictly Presbyterian spirituality owed anything more to fairies and spectres than to Calvin and Knox. In later life, he did recall a couple of his own childish visions, and also adult hallucinations in the fever of smallpox, first of a tea kettle, and then of a waterfall of blood which he had once seen as a theatrical special effect. But his discussion is very much in the context of the search then being made by Walter Scott the writer, and David Brewster the physicist, for naturalistic explanations, such as disturbances of perception, for the 'supernatural'.

The bank job gave Miller financial security at last. He and Lydia Fraser were married on 7 January 1837 by the Revd Stewart. After two days' honeymoon in Elgin, the Millers set up home in the stone house, next to his birthplace, which his father had built. Thus began a companionate marriage of two people well suited to each other. If not rich in the worldly sense, they could at least furnish a parlour, bedroom and kitchen, while he had shelves and a table for his fossils in an attic. To supplement his annual salary of £60, he worked as a freelance author for the periodicals *Border Tales* and *Chambers's Edinburgh Journal*, and carried on his writings in local history, publishing a *Memoir of William Forsyth, Esq. A Scotch Merchant of the Eighteenth Century* (1839). Lydia continued to teach a few girls, and took an evening school for fisher-lads who were at work during the normal schooling day.

In its last years, Miller's bachelor life had enjoyed a pleasant simplicity, as Harriet Taylor recalled:

> To judge from what I saw of him in the earlier years of which I have been speaking, as well as at a later period, I do not think that he had any desire to occupy a different social position than that in which he was born; he had indeed ambition; but it was to be recognised as a man of intellect in the scientific and literary world. I think he was happy in those days; for his wants were few[,] his time much at his own disposal, and he was greatly

respected by all in his native town save those who disliked goodness, and they had felt his power more than once when they set themselves to oppose his minister – and his literary ability had by this time been acknowledged by good judges.[9]

Miller had indeed, as a youth, fantasised about becoming a literary hermit. Lydia recalled him, even after several years of married life, as 'capable of existing with perfect convenience in a cave, one stone serving for table, another for seat, and a plate, knife and fork the whole "plenishing"', and suffering none of Lydia's anxiety to 'put things on a respectable footing'.[10] Although this caveman manqué accepted that such a lifestyle would not do for Lydia, he perhaps never quite entered wholeheartedly into the rituals, responsibilities and expectations of middle-class life:

> To dress better, to live in a better house, to have daintier fare – all these were things to him as if they were not. Indeed, the absence of that ordinary kind of ambition in him might be said to amount to a fault. I might want respectable tables and chairs, and take pleasure in painting and garnishing my house, and he might look upon all this as something to be expected in the feminine gender, and to be tolerated accordingly; but, for himself, he professed just what he felt – to be content with a table, a chair, and a pot, with a little fire in his grate and a little meat to cook on it.
>
> He took pleasure when he could – for that was rare – in sitting with fossil-shelves and book-shelves around him, and with a heap of literary confusion about, which was order to him, and which no hands might touch. And if I came in and sat on his knee and talked to him a little, that was his paradise. But of personal ambition, other than to write something which men would not willingly let die, he had not a single grain.[11]

And there was family to think about. Hugh and Lydia's first daughter, Eliza, was born in 1837. When she died of a fever before her second birthday, Miller was overcome with grief:

> Never again in the course of his life was he thus affected. He was an affectionate father, and some of his children were at times near death; but he

never again lost thus the calmness and dignity, the natural equipoise as it were, of his manhood.[12]

Miller took up mell and chisel for the last time as a mason, and wrought the headstone which still marks Eliza's grave in St Regulus's kirkyard, in the woods towards the South Sutor where he and Lydia had once courted.

NOTES TO CHAPTER 6

1. *Schools*, p. 519.
2. *Schools*, p. 504.
3. L. M. F. F. Miller, 'Mrs Hugh Miller's Journal', p. 463.
4. *Schools*, pp. 507–8.
5. *Schools*, p. 516.
6. Sutherland, E. and M. A. McKenzie Johnston, *Lydia, wife of Hugh Miller of Cromarty* (East Linton, 2002), pp. 166–67.
7. 'Public dinner in Cromarty. Tribute to Mr Hugh Miller', *Inverness Courier*, 15 January 1840.
8. The fossils largely vanished from the second edition of 1850, presumably because they were now dealt with in *The Old Red Sandstone*.
9. Sutherland and McKenzie Johnston, *Lydia*, p. 166.
10. *Life*, vol. 2, p. 219.
11. L. M. F. F. Miller, 'Mrs Hugh Miller's Journal', p. 515.
12. L. M. F. F. Miller, 'Mrs Hugh Miller's Journal', p. 513.

REVEREND PROFESSOR THOMAS CHALMERS

The outstanding preacher and Evangelical, and, with Miller, a key figure in the Disruption which gave rise to the Free Church of Scotland. Calotype by D. O. Hill and R. Adamson.
Original in two-volume album of calotypes presented to the Society of Antiquaries of Scotland by D. O. Hill and others (1851), NMS Library Accession 3250.

(IMAGE © NATIONAL MUSEUMS SCOTLAND)

CHAPTER 7

A plain working man, in rather humble circumstances[1]

AFTER a period of doubt in his younger days, Hugh Miller settled around 1825 into a lasting commitment to Calvinist Presbyterianism. Perhaps his illness had shaken up his thoughts; he was undoubtedly influenced by John Swanson, now studying for the ministry, and by Cromarty's new parish minister, the outstanding preacher Alexander Stewart. Stewart and Miller became good friends, talking as they walked along the shore or as Miller wrought on a gravestone in the kirkyard. Indeed, to visit Cromarty old parish kirk is to gain an insight into Miller. The austere and altarless kirk's T-shaped plan centres the attention of the seated congregation on the pulpit, reflecting the Scottish Presbyterian tradition of minimising formal ritual in favour of long sermons to be carefully considered by the audience.

Under the Presbyterian system enacted by John Knox, Andrew Melville and the other sixteenth-century Reformers, each parish was (and is) run by a 'kirk session', a committee comprising the minister and a number of 'elders' picked from the congregation. Each session sent representatives to the Presbytery of their area, which in turn sent representatives to the annual General Assembly, convened by an elected Moderator. The Church of Scotland regarded Christ as its permanent head, and always resisted the imposition of Government control through a fixed hierarchy of bishops, as in the Church of England. At the local level too, there was much resistance to the right of ecclesiastical patronage whereby lairds could impose their choice of parish minister on a congregation. Many regarded such 'intrusion', as it was called, as offensive in principle. It was an infringement of personal liberty and religious conscience. Parishioners also worried that the laird would choose a minister for social compatibility rather than competence.

All those were important issues in a Scotland which was far more centred on the Kirk than it is today. Knox and the other Scottish Reformers had sought to make Scotland a 'Godly Commonwealth', and Scotland certainly became what has been called a parish state, where the parish bore many responsibilities which we today consider the realm of local and even national government. For instance, the kirk session ran the school and paid the schoolmaster, and might even help students attend university (Presbyterians placed great emphasis on personal study of the Bible and church doctrine, and therefore on literacy). The session also supported the deserving needy and unemployed, exerted basic social discipline on the idle, drunken and disruptive, and obtained paternity support for unmarried mothers.

These activities were supported by voluntary donations and by teinds, which were effectively a land tax on local landowners. Consequently, the laird's power to choose the parish minister was potentially in conflict with those fiscal obligations, and could severely undermine the parish's ability to fulfil its social duties. Landowners could also, and often did, block the revision of parish boundaries, which many thought increasingly urgent in response to the growth of the cities and the rise of new mining and industrial settlements, which made the traditional boundaries increasingly obsolete.

The problem of patronage had caused the breakaway of independent Presbyterian Churches (Secession Church, 1733; Relief Church, 1761), and continued to cause recurrent trouble within the Church of Scotland, with court cases, lockouts, boycotts, and public disorder, in Cromarty as elsewhere. While Miller was growing up, the Church of Scotland became polarised into the so-called 'Evangelicals' and 'Moderates'. 'Evangelical' was, of course, the general term applied to the late eighteenth- and early nineteenth-century religious revival in Britain. But in early nineteenth-century Scotland 'Evangelical' became the name of a party within the Kirk strongly committed to an activist Christianity. Thomas Chalmers, their leader, sought to create the ideal 'Godly Commonwealth' even in the new urban and industrial areas of Scotland, pioneering the techniques of social work in a Glasgow slum parish. The Evangelicals sought reform of patronage and of the parish system. In opposition, the Moderates were happy to live with the status quo as an expression of tolerance, and respected patronage as a right of property.

The Evangelical-dominated General Assembly of 1834 gave male heads of households the right to reject the patron's candidate for their parish minister. The

Earl of Kinnoull's candidate for Auchterarder was rejected by the congregation. The local presbytery therefore refused to ordain him and were sued by the candidate. This key case went in 1839 to the House of Lords, the final court of appeal (in which Scots members were a minority). Lord Brougham (a Scot) and the other judges found in the candidate's favour.

Like many Cromarty burghers before and since, Miller's sentiments were firmly Evangelical, and he was so outraged by Brougham's judgement, and the sneering tone in which it was expressed, that he wrote an angry article, 'Letter from one of the Scotch people to the Right Hon. Lord Brougham & Vaux, on the opinions expressed by His Lordship in the Auchterarder Case'. He started with the ironic deference and brisk egalitarianism appropriate for a matter which was as much about aristocratic presumption as religious liberty:

> MY LORD, – I am a plain working man, in rather humble circumstances, a native of the north of Scotland, and a member of the Established Church.[2]

He noted that Brougham, who had had a crucial role in parliamentary reform, held people to be worthy of the resulting vote, but

> ... unworthy of being emancipated from the thraldom of a degrading law, the remnant of a barbarous code, which conveys them over by thousands and miles square to the charge of patronage-courting clergymen, practically unacquainted with the religion they profess to teach? Surely the people of Scotland are not so changed but that they know at least as much of the doctrines of the New Testament as of the principles of civil government, and of the requisites of a gospel minister, as of the qualifications of a Member of Parliament![3]

The Church of Scotland's members did not

> ... look to the high places of the earth when we address ourselves to its adorable Head. The Earl of Kinnoull is not the Church, nor any of the other patrons of Scotland. Why, then, are these men suffered to exercise, and that so exclusively, one of the Church's most sacred privileges? You tell us of 'existing institutions, vested rights, positive interests'. Do we not know that

the slaveholders, who have so long and so stubbornly withstood your Lordship's truly noble appeals in behalf of the African bondsmen, have been employing an exactly similar language for the last fifty years; and that the onward progress of man to the high place which God has willed him to occupy has been impeded at every step by 'existing institutions, vested rights, positive interests'?[4]

And as for the gentry,

> The Church has offended many of her noblest and wealthiest, it is said, and they are flying from her in crowds. Well, what matters it? – let the chaff fly![5]

Robert Paul, the Commercial Bank's general manager, happened to be an eminent Edinburgh Evangelical, and Miller naturally sent him the manuscript for its successful publication in pamphlet form. But by chance Paul was one of a group of Evangelical opponents of patronage who were so frustrated by the lack of support from newspapers, whether conservative or free-market liberal, that they planned to start their own Edinburgh newspaper, 'without supporting any of the old parties in the State ... as Liberal in its politics as in its Churchmanship'.[6] However, '[a]ll the ready-made editors of the kingdom, if I may so speak, had declared against them',[7] and it was only when Miller's manuscript fortuitously arrived that they found their editor. Miller accepted the job, despite misgivings about his abilities, and worries about involvement in controversy:

> ... the editorship of a Non-Intrusion newspaper involved, as a portion of its duties, war with all the world. I held, besides – not aware how very much the spur of necessity quickens production – that its twice-a-week demands would fully occupy all my time, and that I would have to resign, in consequence, my favourite pursuit – geology. I had once hoped too – though of late years the hope had been becoming faint – to leave some little mark behind me in the literature of my country; but the last remains of the expectation had now to be resigned. The newspaper editor writes in sand when the flood is coming in. If he but succeed in influencing opinion for the present, he must be content to be forgotten in the future. But believing the cause to be a good one, I prepared for a life of strife, toil, and comparative

obscurity. In counting the cost, I very considerably exaggerated it; but I trust I may say that, in all honesty, and with no sinister aim, or prospect of worldly advantage, I *did* count it, and fairly undertook to make the full sacrifice which the cause demanded.[8]

On 8 January 1840, Miller's friends held a public dinner in his honour at Cromarty at which, much to his gratification, Uncle Sandy was present. Here Miller was given a silver breakfast service by 'a kind and numerous circle of friends, of all shades of politics and both sides of the Church' in 'testimony of their admiration of his talents as a writer, and of their respect and regard for him as a member of Society'.[9]

Miller took ship for Edinburgh the next morning.

NOTES TO CHAPTER 7

1. 'Letter to Lord Brougham', pp. 1–22 in Miller, H. *The Headship of Christ, and the rights of the Christian people* (Edinburgh, 1889), p. 2.
2. 'Letter to Lord Brougham', pp. 1–22 in *Headship*, p. 2.
3. 'Letter to Lord Brougham', pp. 1–22 in *Headship*, p. 3.
4. 'Letter to Lord Brougham', pp. 1–22 in *Headship*, p. 8; 'high places' alludes to the biblical places of hilltop and often idolatrous sacrifice, e.g. Leviticus 26:29–31.
5. 'Letter to Lord Brougham', pp. 1–22 in *Headship*, p. 21.
6. *Schools*, p. 549.
7. *Schools*, p. 550.
8. *Schools*, p. 554.
9. *Schools*, p. 560; *Inverness Courier*, 8 and 15 January 1840.

DR JOHN MALCOLMSON

Malcolmson was an army doctor in India and a keen geologist,
who played a crucial role in linking Hugh Miller to other geologists.
Original picture in Falconer Museum, Forres, 1978-316.

(IMAGE © MORAY COUNCIL MUSEUMS SERVICE)

CHAPTER 8

Among the remains of a different creation[1]

MILLER'S Cromarty fossil fishes had puzzled him ever since he found them in 1830. But, as his isolation from other geologists slowly broke down over the 1830s, he came to realise that he had made new discoveries which established his position in the geological community.

On one level, during the 1830s, Miller simply continued to exploit the little bay, temptingly close to Cromarty (and today protected as part of the Rosemarkie to Shandwick Site of Special Scientific Interest):

> ... when fatigued with my calculations in the bank, I used to find it delightful relaxation to lay open its fish by scores, and to study their peculiarities as exhibited in their various states of keeping, until I at length became able to determine their several genera and species from even the minutest fragments. The number of ichthyolites which that deposit of itself furnished – a patch little more than forty yards square – seemed altogether astonishing: it supplied me with specimens at almost every visit, for ten years together; nor, though, after I left Cromarty for Edinburgh, it was often explored by geologic tourists, and by a few cultivators of science in the place, was it wholly exhausted for ten years more.[2]

Lydia recalled:

> While breaking open his specimens or thinking over them at his work, he was intensely happy – the happiness of an unruffled spirit at peace with God and all men. ...[3]

Miller also searched for more sites bearing fossil fish in the Old Red Sandstone around Cromarty, sometimes during a cruise with Lydia in his little yawl to picnic and fish along the coast. But, because of his post at the bank, he could only go collecting on summer evenings and early mornings, and Saturday afternoons (although not Sundays). Years later, he ruefully recalled being

> … somewhat in the circumstances of a tolerably lively beetle stuck on a pin, that, though able, with a little exertion, to spin round its centre, is yet wholly unable to quit it.[4]

On another level, however, for much of the 1830s Miller made little progress in understanding his finds. He did not know what to call his 'several genera and species', which is simply to say that he did not know what they were, in an age which was preoccupied above all with exploring, sorting and naming the natural diversity of creation. To begin with, he had little more than newspaper articles and an occasional, and often outdated, book to guide him. But in 1834 the Anderson brothers of Inverness published their *Guide to the Highlands and Islands of Scotland*. Its geological appendix gave Miller a basic summary of the region's geology, as it was then understood. However, the age of his fossils was still unclear, and the book did not help much with Miller's particular problems. Nor were the fossils themselves particularly helpful, being often disrupted by the vagaries of decay and burial, and squashed by the weight of overlying sediment. Indeed, some were found only as broken fragments. Miller faced three-dimensional jigsaws of the most difficult kind, without even being able to fit the parts physically together. To help, he sometimes painted letters or numbers on the bony plates when trying to sort them out.

Miller sought to improve his knowledge of living fishes by obtaining examples from Cromarty fishermen, but this took him little further, for he could not even relate his fossils to living families:

> It was in vain I examined every species of fish caught by the fishermen of the place, from the dog-fish and the skate to the herring and the mackerel. I could find in our recent fishes no such scales of enamelled bone as those which had covered the *Dipterians* and the *Celacanths*; and no such plate-encased animals as the various species of *Coccosteus* or *Pterichthys*. On the

other hand, with the exception of a double line of vertebral processes in the *Coccosteus*, I could find in the ancient fishes no internal skeleton: they had apparently worn all their bones outside, where the crustaceans wear their shells, and were furnished inside with but frameworks of perishable cartilage. It seemed somewhat strange, too, that the geologists who occasionally came my way – some of them men of eminence – seemed to know even less about my Old Red fishes and their peculiarities of structure, than I did myself. ... I at length came to find that I had got into a *terra incognita* in the geological field ... .[5]

Some of Miller's fishes were indeed new to science. Together with similar finds being made elsewhere, they were amongst the oldest fossil vertebrates then known, as Miller would learn, now that his isolation was breaking down. For instance, on the other side of the Moray Firth Miller knew, from his attempts to sell his book of poetry, the Elgin bookseller Isaac Forsyth, one of a small group of people who, in October 1836, founded the Elgin and Morayshire Scientific Association.[6] When Miller and Lydia came to Elgin for their honeymoon, it was Forsyth who showed them round; thus Miller got to know Patrick Duff, town clerk of Elgin and keen geologist. Thereafter Miller and Duff maintained a lively correspondence for a while, discussing their respective local fossil fishes.

Miller's account of the fishes in *Scenes and Legends* attracted the attention of some geologists, such as the Revd Professor John Fleming of King's College, Aberdeen, and notably Dr John Malcolmson, an army doctor on sick leave from India. In late 1837 Malcolmson was visiting his mother in Forres and seized the opportunity to come over to Cromarty. He transformed Miller's geology with his friendly and knowledgeable enthusiasm. Lydia's reaction to his visits, however, reflected the apparent strangeness of Miller's newly energised hobby:

The deciphering of the *Pterichthys*, the putting together of bone and plate as they happened to turn up, was the great work of the evening. ... I found these endless discussions rather trying. I didn't see just then how these dead bones were to live. I sat at my work listening, wishing often that there would be a change of topic; but the interest of the two gentlemen was unwearied, their discussions unflagging.[7]

Malcolmson came over as often as he could. He helped Miller to sort out some of his problems and realised the confusion between the fishes later to be named *Coccosteus* and *Pterichthys*. In time, with Malcolmson's help, the whole animals eventually emerged:

> ... every new specimen that turned up furnishing a key for some part previously unknown – until at length, after many an abortive effort, the creatures rose up before me in their strange, unwonted proportions, as they had lived, untold ages before, in the primaeval seas.[8]

Malcolmson would also have brought Miller fully up to date on mid-1830s geology. This was now an empirical science, in which fossils were important: first, as markers for rocks of different ages, enabling geologists to discern the relative ages of strata, and to map the country's rocks; and second, as evidence in themselves for past life. What the fossil record had appeared to show was a long history of life on Earth, periodically marked by massive catastrophes, each of which wiped out all life, and was followed by a new divine creation. Hence Miller's description of Old Red fishes as 'the remains of a different creation'[9] from his fossils of the Lias at Eathie – and different again, of course, from present-day life.

> The huge sauroid fish was succeeded by the equally huge reptile, – the reptile by the bird, – the bird by the marsupial quadruped; and at length, after races higher in the scale of instinct had taken precedence, in succession, the one of the other, the sagacious elephant appeared, as the lord of that latest creation which immediately preceded our own.[10]

Even then, it was realised that the transitions across supposed catastrophes were more gradual than initially supposed, with some living forms surviving into the new period. But, even so, it did seem that those successive catastrophes and creations had modified the Earth's climate and life. This gave in, the end, a habitat suitable for humanity, which was itself duly specially created after the most recent catastrophe, and would face the next and last geological catastrophe in the Last Judgement.

Malcolmson's impact on Miller also came about through his wider contacts, for he was a Fellow of the Geological Society of London, the prime British venue for the science. Malcolmson took some of Miller's fishes when he travelled to London in February 1838, and asked him to send more down to London, to go via Roderick Murchison, Scottish geologist and key figure in the Society, to the Swiss Professor Louis Agassiz. Crucially, Agassiz was then the world expert on fossil fishes, and author of the influential book *Poissons fossiles* [*Fossil Fishes*]. He would eventually publish one of Miller's fossils in his book *Poissons fossiles du vieux grès rouge* [*Fossil Fishes of the Old Red Sandstone*] of 1844–45. But even before that, Agassiz and Murchison noted Miller's discoveries in his absence at the Geological Section of the British Association for the Advancement of Science, then the premier national forum for science, at its meeting in Glasgow in 1840. These fishes were considered something of a sensation – particularly Miller's 'singularly formed animal with lateral wing-like processes, the *Pterichthys*'.[11] Agassiz, on the spot, named the most prominent species *Pterichthys milleri*,[12] reflecting the honour then due to the discoverer of a fossil whose work had made it known to the scientific community. Miller, through his researches and Agassiz's work, thus gained the respect of the scientific community in a new science then rising to the height of fashion in Britain.

Miller's finds, then and later, were indeed a substantial contribution to the diversity and geological distribution of the Old Red Sandstone fishes, of Devonian age. His key Cromarty finds, the two genera *Pterichthys* and *Coccosteus*, are today recognised as members of the placoderms, a group of armoured fishes. Certain of Miller's specimens of *Pterichthys milleri* were amongst the first of many fossils in his collection, now in National Museums Scotland, to attain the status of type specimens, upon which the names of genera and species are defined, crucially enabling scientists to establish the language of their discussions. They are amongst National Museums Scotland's most precious objects.

Then and later, Miller made useful observations on the anatomy of the Old Red Sandstone fishes, correcting even the great Agassiz. Thomas Henry Huxley, better known as the combative defender of Darwin's evolutionary ideas, would later remark:

> The more I study the fishes of the 'Old Red', the more am I struck with the patience and sagacity manifested in Hugh Miller's researches, and by the natural insight which in his case seems to have supplied the place of special anatomical knowledge.[13]

Here, of course, Huxley was praising Miller's descriptive work on the basic structure of his fossils; evolutionary interpretations were a quite different matter on which the two would have disagreed, as will later be seen.

Miller was thus making his first mark in geology. In one sense, geology is simply what one does when studying rocks, like Miller collecting on the Cromarty beach. But, in another sense, geology, like all science, is a collective activity: where in it was Miller to fit?

Today, someone wanting to do geology must effectively decide whether to be a spare-time amateur, or a salaried professional such as a state survey geologist or a university academic. That distinction seems fundamental to us, but in the 1830s it was almost meaningless: geology was so new that it had yet to form the institutional and career structures familiar today. Miller did, it is true, make some money from geology, by writing about it (but not from selling fossils, as far as is known) – but this would account for only a tiny fraction of his literary earnings before 1841, and even after that, only a part of them. Few university professors, and a limited number of popular writers, peripatetic lecturers, curators, surveyors and engineers, made a substantial part of their living from this new science.

Today, another crucial distinction in geology is whether one writes formal scientific papers. Miller, despite his reputation and his real skills as a geologist, published little in the way of formal scientific papers, even after 1840. But this simply reflects the varied reasons and ways of doing geology in 1830s Britain. For most people involved, geology was a purely voluntary activity. Geologists found the niches they liked best. Some liked to write detailed classificatory monographs; some explored and mapped the landscape; some collected fossils to reveal the past of their locality; some liked to collect or draw fossils and minerals for their beauty; some wrote about them; some helped organise societies and museums for the local public benefit, for education and civic pride, and to build social bridges, as well as for their own private, religious and social satisfactions; and some just followed the fashion of the day.[14]

Miller had few options stuck out on his own at Cromarty, although one can imagine he would have joined the Elgin society if he had lived on the other side of the Moray Firth. Nevertheless, he found his own niches in the science as soon as, and indeed perhaps even before, he made full contact. It seems that Miller simply picked what he liked doing. It is possible to imagine that he was too busy

Above

CROMARTY FROM THE EAST
Print by John Clark, 1820s

An important trading town thanks to the superb natural harbour of the Cromarty Firth. Fishing boats are beached in front of the town; trading vessels lie further up the Firth.

CROMARTY COURTHOUSE.
LICENSOR WWW.SCRAN.AC.UK

Left, above

IN THE FISHERTOWN, CROMARTY

Taken about 1900, but no doubt much as in Miller's time – including the scents of fish and bait.

(ST ANDREWS UNIVERSITY LIBRARY. LICENSOR WWW.SCRAN.AC.UK)

Left, below

HUGH MILLER'S COTTAGE

The Birthplace Cottage, built by Miller's great-grandfather, now in the care of the National Trust for Scotland.

DR M. A. TAYLOR

Above

EDINBURGH STONEMASONS WORKING ON THE SCOTT MONUMENT

David Octavius Hill and Robert Adamson, calotype, c. 1845. Original in album of calotypes presented to the Society of Antiquaries of Scotland by Hill and others (1851). NMS Library accession 3250.

IMAGE © NATIONAL MUSEUMS SCOTLAND

Left, above

THE HOUSE THAT MILLER HELPED BUILD

Niddrie House near Edinburgh, now long demolished. From J. Small: *The castles and mansions of the Lothians* (Edinburgh, 1883).

IMAGE © NATIONAL MUSEUMS SCOTLAND

Left, below

THE HOUSE IN WHICH MILLER LIVED

A pair of Midlothian farm cottages typical of the one-roomed dwelling with an earthen floor where Miller and a colleague lodged with an elderly couple. From G. Robertson: *General view of the agriculture of the county of Mid-Lothian* (London, 1795).

IMAGE © NATIONAL MUSEUMS SCOTLAND

Above

PRINCES STREET WITH THE COMMENCEMENT OF THE BUILDING OF THE ROYAL INSTITUTION

Alexander Nasmyth, oil painting, 1825 (detail)

The masons are working on the Institution, now the Royal Scottish Academy: part of the building boom in Edinburgh which drew Miller in search of work.

NATIONAL GALLERIES OF SCOTLAND

Right

HUGH MILLER, 1802–1856. GEOLOGIST AND AUTHOR

David Octavius Hill and Robert Adamson, calotype, 1843 (detail)

The bunnet (in English, bonnet) was the soft hat of the outdoor Scotsman. This calotype is by the pioneer Edinburgh photographers David Octavius Hill and Robert Adamson.

SCOTTISH NATIONAL PORTRAIT GALLERY

Above
ONE OF MILLER'S WORKPLACES
The parish kirkyard at Nigg, Easter Ross, just across the Firth from Cromarty.
DR M. A. TAYLOR

Left, above
JAMES WRIGHT'S GRAVESTONE IN CROMARTY KIRKYARD
Miller carved this gravestone for his uncle James Wright in his plain but pleasing style.
DR M. A. TAYLOR

Left, below
SUNDIAL
Miller carved this sundial for his uncles. Now in the garden of Hugh Miller's Birthplace Cottage, Cromarty.
DR M. A. TAYLOR

Opposite page, top left
MILLER HOUSE
Built next to the Cottage (whose gable end is seen here) by Miller's father, rented out during Miller's childhood, and finally occupied by Hugh and Lydia Miller when they married in 1837.
(DR M. A. TAYLOR)

Opposite page, top right
EATHIE BURN AS IT FLOWS DOWN TO THE SEA
Miller recorded in *Scenes and Legends* that the last fairies in Scotland were seen here.
(FROM BOOK OF PHOTOGRAPHS PUBLISHED BY JOHN BAIN, DRAPER AND CLOTHIER [CROMARTY, 1900])

Left
ROBERT CARRUTHERS
David Octavius Hill and Robert Adamson, calotype (detail)
Editor of the *Inverness Courier*, Carruthers gave Miller his first regular journalistic work as Cromarty correspondent.
SCOTTISH NATIONAL PORTRAIT GALLERY

Right
ELIZA MILLER'S GRAVE
IN ST REGULUS'S KIRKYARD
The last gravestone Miller ever carved, for Lydia's and his first child. It shows the scalloped edge often seen on his work.
DR HELEN HANDOLL

Above
MILLER'S FIRST GEOLOGICAL STAMPING GROUND
Miller's interest in fossils was aroused when he was quarrying nearby and a workmate told him of fossils to be seen in the rocks on the shore here at Eathie.
DR LYALL ANDERSON

Left, above
AN EATHIE AMMONITE
Collected by Miller. A spiral shellfish related to modern squids and nautilus. *Xenostephanus* aff. *anceps*, specimen 11 cm long. NMS.G.1859.33.3957.
IMAGE © NATIONAL MUSEUMS SCOTLAND

Left, middle
A 'THUNDERBOLT'
In fact, a belemnite: the fossil guard or internal shell of a squid-like animal. Collected by Miller from Eathie. *Cylindroteuthis spicularis*, 19 cm long. NMS.G.1859.33.4026.
IMAGE © NATIONAL MUSEUMS SCOTLAND

Left, below
EATHIE'S ABUNDANCE OF FOSSILS
Small ammonites crowding what was once the Jurassic sea-floor. *Amoeboceras* cf. *nathorsti*, specimen 17 cm long. NMS.G.1859.33.3989.
IMAGE © NATIONAL MUSEUMS SCOTLAND

Above
THE SITE OF MILLER'S CLASSIC DISCOVERY OF OLD RED SANDSTONE FOSSIL FISHES

Showing the headlands of the Sutors in the distance, where the Cromarty Firth opens into the Moray Firth.

DR M. A. TAYLOR

Left
THE LATE MR HUGH MILLER, AUTHOR OF THE OLD RED SANDSTONE

Amelia Paton Hill, 1860s

Miller has just found a specimen of the fossil fish *Pterichthys milleri* by splitting a nodule on Cromarty beach. The other half of the nodule is lying at his feet. Statue about 1.2 m high.
C. Baile de Laperrière (ed.): *The Royal Scottish Academy exhibitors 1826–1990. A dictionary of artists and their work in the Annual Exhibitions of The Royal Scottish Academy* (Calne, 1991).

NMS.A.1887.735.

IMAGE © NATIONAL MUSEUMS SCOTLAND

Above
CROMARTY PARISH KIRK
Miller's family church. Now the East Kirk, cared for by the Scottish Redundant Churches Trust.
DR M. A. TAYLOR

Left
CROMARTY PARISH KIRK INTERIOR
Showing the middle of three wings of this T-shaped church focussed on the preacher's pulpit (just out of view to the right), as befits a Presbyterian kirk. Miller's family rented part of the 'table seat' at the front of the gallery: the one with 1741 painted on the front, on the left of the picture.
RCAHMS/SCRAN

to be a research geologist on a grand scale. No doubt, too, he preferred being out collecting fossils. But Miller also found his own very special niche, a popularising and moralising role in writing about geology which also met his aim of literary achievement. He certainly started to use his hobby to generate useful material for journalism even before he left Cromarty in 1840. His first publication on geology (other than the material in *Scenes and Legends*) was a double piece on his early 'gropings' in *Chambers's Edinburgh Journal* in 1838. Moreover, at his farewell dinner, Miller recalled his literary ambitions to make Cromarty known; once he had dealt with local history and tradition, his attention had turned to Cromarty's natural history –

> ... I have set myself to exhibit it as a locality, in which the naturalist – a White of Selborne, for instance – might have delighted; and I straightway found that I had travelled into a new district. Objects, before unnoticed, or but slightly regarded, rose into interest. Even the spiky leaves and light florets of the thirty or forty sorts of the humble family of the grasses, which I met with in my short walks; every insect that enjoyed itself on the breeze or the stream, grew into beauty and importance. Still more was I interested when, passing from the present creation, I found the locality a rich museum of the remains of former creations.[15]

Evidently Miller abbreviated his chronology for rhetorical effect: for instance, *Scenes and Legends* appeared well after Miller had seriously begun geologising. But this does suggest that Miller might have started writing about geology even before the 1838 articles. And his ambitions were certainly growing at the end of 1838, if they hadn't already, for that was when he told Duff of his intention to draw up an account of the geology of Cromarty over the summer of 1839.[16]

In any case, when at his newspaper editor's desk in 1840, Miller apparently spotted the opportunity posed by the forthcoming meeting of the British Association at Glasgow. Possibly drawing upon unpublished Cromarty writings of the kind he mentioned to Duff, he published a series of articles on 'The Old Red Sandstone' in his newspaper over September and October, the first few of which coincided with the meeting. Murchison publicly praised Miller's style as 'so beautiful and poetical as to throw plain geologists like himself into the shade',[17] and William Buckland of Oxford University, himself an important writer on geology, lauded Miller's literary powers. Those comments were not widely reported at the time, but Miller must have been encouraged. He updated his articles (with one of 15 April that had recycled his first 'gropings' article and another of 16 December) and added much more content, heavily reorganising and rewriting all this into what was effectively a new book, *The Old Red Sandstone*, published in 1841. This completed the process his finds had begun of making Miller's name in geology. It is, on one level, a description of the geology of Cromarty, with diversions into scenery, history and folklore, added to a wide-ranging review of the Old Red Sandstone fishes and their scientific and philosophical implications. But it is also a delightful autobiographical memoir and work of literature. It was enormously popular, and deservedly so.[18]

At the start of the 1840s, therefore, Miller (his own observations still largely unpublished) would have seemed, from one point of view, yet another of the many provincial collectors exploring beaches and quarries all over Britain, and yielding up their fossils to the pronouncements of the metropolitan scientists and their international contacts. He had thereby added a pebble, so to speak, to the growing cairn of geological knowledge – and one of which he could be proud. Yet this was merely the beginning of his geological work, in more than one sense, for Miller, above all else, retained his literary ambitions. He would select another, and rather different, path to that chosen by many of his new metropolitan contacts. He soon made his mark on this path with *The Old Red Sandstone*, presaging many of his later geological writings; in this, as with so much else, Miller reached back to his Cromarty origins.

NOTES TO CHAPTER 8

1. *Old Red Sandstone*, p. 131.
2. *Schools*, p. 526.
3. L. M. F. F. Miller, 'Mrs Hugh Miller's Journal', p. 515.
4. *Schools*, p. 530.
5. *Schools*, pp. 526–27.
6. Today the Moray Society, in Elgin Museum.
7. L. M. F. F. Miller, 'Mrs Hugh Miller's Journal', p. 464; here, 'work' is sewing.
8. *Schools*, pp. 527–28.
9. *Old Red Sandstone*, p. 131.
10. *Old Red Sandstone*, pp. 270–71; the 'sauroid' fish was presumably one of the giant rhizodontids

(predatory freshwater fishes up to several metres long) of the Coal Measures, while the 'elephant' would include mammoths and mastodons.

11. R. I. Murchison, 'Fishes of the Old Red Sandstone', *Annual Reports of the British Association for the Advancement of Science, for 1840* (1841), Sectional Reports, p. 99.
12. Now *Pterichthyodes milleri*. *Pterichthys* is from the Greek for 'wing' and 'fish'.
13. Anon., *Centenary*, p. 60. Not located in Huxley's published writings, so perhaps in a letter to the Miller family or Hugh the younger. Evidently related to the research published as e.g. T. H. Huxley, 'Preliminary essay upon the systematic arrangement of the fishes of the Devonian Epoch', *Memoirs of the Geological Survey of the United Kingdom: figures and descriptions illustrative of British organic remains*, Decade **10**, 1–40 (London, 1861), in which Huxley praises Miller's work. See M. A. Taylor, 'Three memoirs of Hugh Miller (1802–1856) by his son Hugh Miller FGS', *Archives of Natural History* **46**, 113–18 (2019).
14. S. J. Knell, *The Culture of English Geology, 1815–1851* (Aldershot, 2000); Knell explicitly restricts his analysis to England, but it seems to me to be applicable to Scotland, certainly as a first approximation.
15. 'Public dinner in Cromarty. Tribute to Mr Hugh Miller', *Inverness Courier*, 15 January 1840. *The natural history and antiquities of Selborne* (1789), the classic study of a single parish, was written by the Revd Gilbert White of Selborne in Hampshire.
16. Miller to Duff, 15 December 1838, Elgin Museum Geology Letter G1/2.
17. Anon., *Centenary*, p. 57.
18. See also new analysis by R. O'Connor and M. A. Taylor, 'A critical study of Hugh Miller's *The Old Red Sandstone*, 1838–1920', in Hugh Miller, *The Old Red Sandstone* (reprint of 1841 edition, Edinburgh, forthcoming).

CHAPTER 9

Strife, toil, and comparative obscurity[1]

THE first issue of the *Witness* appeared on 15 January 1840. The *Witness* was no ecclesiastical magazine or church news-sheet. It was a mainstream newspaper, typical of Scottish newspapers in that it was published twice a week, in this case on Wednesday and Saturday, with later news in a second, afternoon, edition. A copy usually comprised one large sheet of paper folded into four pages of finely printed text. Its price was also typical at fourpence halfpenny a copy, which was not cheap (but each one would no doubt be read by several people). This was due partly to the so-called 'taxes on knowledge' levied on paper, on advertisements, and on each copy of the newspaper (although this last Stamp Tax at least included 'free' postage to distant subscribers).

The *Witness* was initially published by John Johnstone of Hunter Square in Edinburgh. The editorial offices were at 297 High Street, Edinburgh, a few moment's scamper for a printer's boy past the High Kirk of St Giles, down steep Old Fishmarket Close, and along the Cowgate to the printing shop at the foot of Horse Wynd (now the lower part of Guthrie Street). The initial proprietors, the founding subscribers, together put up £1000, apparently passed to Johnstone to buy printing equipment. The printer, Robert Fairly, added a further sum, thereby becoming a partner and co-publisher, and apparently acted as a general manager.[2] Miller himself started as an employee at an annual salary of £200, heading a tiny permanent journalistic staff, perhaps – at least to begin with – just the usual editor, sub-editor and reporter of 'the lower orders of Scotch newspapers'.[3] However, they did not have to fill all four pages with their own writings; some space was devoted to advertisements, while much news was simply taken from other newspapers, such as foreign affairs from the London *Times*.

James Mackenzie, one of Miller's assistants, recalled the Revd William

Cunningham, a shareholder and Miller's friend, visiting the newspaper in those early days: he

> ... would often step up our dingy stair – not unfrequently would have to spend an hour touching up an article, to make it legible for the printer. He has shared our coffee on publishing nights, when we manufactured it in our gigantic coffee-pot, – Hugh Miller, meanwhile, contriving to toast cheese by the help of the fire-shovel. It was a cheery time, in spite of overhanging responsibilities.[4]

The *Witness* soon gained a circulation varying between two and three thousand copies, making it the main contender with the *Scotsman* to become the Edinburgh paper with the largest circulation. Among Edinburgh newspapers, its special character naturally arose partly from its Evangelical (and, later, Free Church) sympathies, especially at Assembly time, when the paper printed much Kirk business. However, its editorial independence was even then evident. Moreover, the more substantial feature articles, many written by Miller himself, gave the newspaper a further special character with something of the weighty flavour of the critical monthlies and reviews of the day.

Miller had left Lydia in Cromarty with their newborn daughter Harriet, but the family was soon reunited in Edinburgh, where the Millers rented 5 Sylvan Place, a short walk to the office across the grassy park of the Meadows. Today the house lies on the north fringe of the suburbs of Sciennes, Marchmont and Grange. But in 1840 the area had a semi-rural feel, with gentlemen's villas, cottages, dairies and market gardens merging into the open countryside and the hills of Blackford and Braid beyond. It is easy to suggest good practical reasons for the location: economy, peace and quiet for Miller to work, and the prevailing wind which kept Edinburgh's pollution mostly downwind of his damaged lungs. But the location also hints at someone who remained a country boy at heart.

Harriet Ross Taylor recalled life there in 1840, when she came to stay:

> The house in Sylvan Place was pleasantly situated; for from the windows in front we saw Arthur's Seat, and from the drawing-room windows an extent of the richly green meadows with fine trees here and there – much built upon now and different. Mrs Miller furnished the other rooms of the house, but

did not find it convenient at that time to fit up the drawing-room suitably; a few things however were put into it and we found it a pleasant sitting-room. Books, which we had brought from Cromarty, and many new ones which came to Mr Miller as the editor of a newspaper were piled up around the walls; and at the foot of the room which was not a small one, his desk was placed. There he wrote all the long day and far into the night except on Wednesday and Saturday afternoons when he took long walks into the country. Mrs Miller and I sat either working, reading, or writing, and took good care never to disturb him. He never sat, but walked up and down, repeating his sentences until they were moulded to his liking; occasionally coming up to his wife and saying, 'Do you think this is the best way to put it?' and when satisfied went to his desk and wrote. At meals that on which he was writing was the subject of conversation, for we were generally alone and often a book was laid on the table out of which something was read.[5]

Lydia ran the growing household while her husband was hard at work: their first son, William, was born in 1842, and their second daughter, Bessie, in 1845. Miller's salary was eventually doubled as the paper's circulation grew, which may be why, some time before October 1843, the family moved over the Meadows to 16 Archibald Place (today, part of the site of the Lauriston Building of the former Royal Infirmary).[6]

Miller could never neglect the cyclical workload of producing the newspaper, and the eternal search for copy, much of which he wrote as editorials, reportage and features. Once various lectures, reviews, poems and other pieces are also added, Miller produced something like 10,000 words a week on average. He added to his own burden by insisting on producing his articles as complete, finished essays, which preserved his characteristic diction, as if to be read aloud at the Victorian fireside – as no doubt they often were. Thomas Guthrie, an Evangelical minister and another proprietor of the *Witness*, noted Miller's elephantine memory for books and poems he had read, quotations from which sprinkle his writing appositely (if often silently changed to suit the cadences of the surrounding text, as was then common). He recalled:

Much of Miller's power lay in the way the subject on which he was to write took entire possession of his mind. For the time being, he concentrated his whole faculties and feelings on it; so that, if we met a day or two before the appearance of any remarkable article … I could generally guess what was to be the subject of his discussion, or who was to be the object of his attack. From whatever point it started, the conversation – before we were done – came round to that; and, in a day or two, the public were reading in the columns of the *Witness* very much of what I had previously heard from his own lips. The subject took possession of him, rather than he of the subject.[7]

There are, to be sure, traces of the journalistic life in some of Miller's writings: recycled themes, pieces squeezed for space or for time, and here and there a lack of Miller's characteristic polish, while some pieces were on topics of only fleeting interest. But the astonishing quality and variety of his articles, many good enough to find new leases of life as pamphlets and in books, conceals the pressure under which they were created. In those articles, Miller was not just an editor, but also a major commentator and critic.

Miller's official biographer, Peter Bayne, drawing upon Lydia's recollections, described Miller thus, presumably on days when the newspaper was not going to press and he could work at home:

> … he modelled his newspaper essays, as he modelled the chapters of his books, on the productions of his beloved Addison and Goldsmith, rather than on those of the 'eminent hands' whose slashing leaders have made their reputation on the London press. It was his habit to fix upon his subject a few days, or even longer, before the article was to appear, and nothing pleased him better than to have Mrs Miller as volunteer antagonist, to maintain against him, at the supper table, the thesis he proposed to controvert. Supper was his favourite meal. At breakfast he hardly tasted food, a cup of coffee and crumb of bread being the limit of his wants. After working at his desk in the early part of the day, he would walk out, make his way into the country, saunter about the hills of Braid or Arthur-seat, with his eye on the plants and land shells and geological sections, or explore for the thousandth time the Musselburgh shore or the Granton quarries. He never clearly admitted the canonical authority of the dinner hour. He expected something warm to be kept ready for him; but if the day was particularly favourable, or if a storm had strewn the coast with the treasures of the deep sea, or if some new

phenomenon struck him in connection with the raised beach at Leith and required interpreting and thinking out, or if he met with a brother naturalist and got into talk, the shades of evening would be falling thick before he again crossed his threshold. Even at that hour he had little appetite. It was not until his brain, obeying what his habits of night-study had made an irresistible law for him, awoke in its fervour about ten o'clock, that he showed a keen inclination for food. Porter or ale, with some kind of dried fish or preserved meat, formed his favourite supper. On these occasions he conversed with great freedom, and found it both pleasant and profitable to have his views and arguments vigorously controverted. There can be no doubt that the extraordinary success of many of his articles ... was due, in a considerable degree, to his having beaten over the ground with Mrs Miller.[8]

As well as an intelligent foil to Hugh's ideas, Lydia was an author in her own right, although the conventions that then bound respectable women limited her to book reviews, romantic novels, and 'improving' books for children. Usually under the pen name of 'Harriet Myrtle', Lydia from 1845 onwards wrote a number of children's books in the moralising Victorian vein, and a romantic novel, *Passages in the Life of an English Heiress or Recollections of Disruption Times in Scotland* (1847). This novel aimed to explain Kirk politics to the English, and like many a good romance it had a naïve heroine, an immoral laird, and a happy ending. Its only known review was in the *Witness*, almost certainly written by Hugh.

This combination of quantity and quality placed a heavy workload on Miller. A family friend, Marion Wood, recalled Miller 'referring to the change from labour with the chisel in the open air to labour at the desk[;] he said the last was far harder than the first, and that the change had come too late to him: "It was too late ere I was caught."'[9]

Cunningham's biographers recalled that Miller

... wrote slowly, and made laborious corrections, speaking out to himself as he wrote, and trying every sentence upon his ear, as a money-changer weighs a piece of gold on his practised finger-tip. 'You must write very easily,' said a complimentary gentleman to him one day. 'Let me tell you, sir,' was the reply, 'that it takes a good deal of hard labour to make writing look easy.'

The publishing days of the *Witness* were Wednesday and Saturday. Each of these was preceded by labour in writing so severe that Miller used to say, 'I can never remember the names of my fossils on publication days till about tea-time, when they begin to come back to me, reappearing to memory like letters written in invisible ink when you hold the paper to the fire'.[10]

It is hardly surprising that Miller took several weeks' annual vacation, often combining a visit to his mother and other relatives in Cromarty with the investigation of some fossil-bearing part of Scotland, gathering material for the *Witness* and for a book on Scottish geology. He seems to have done this mostly alone, no doubt because such trips were unsuitable for young families.

Hugh Miller was now established in his ultimate career. But his literary achievement was only made possible by Lydia being accommodating to his needs. She was a ready foil and a constant housekeeper. Even then, the various locations of the family's houses suggest his unwillingness to engage fully in urban life and to renounce the rural world whence he came.

NOTES TO CHAPTER 9

1. *Schools*, p. 554.
2. *Life*, vol. 2, pp. 203, 269–70; it is unclear whether Johnstone was a partner, and his original commitment was for only one year. See Rainy and Mackenzie, *Cunningham*, p. 135.
3. *Life*, vol. 2, pp. 272, 280, 288; the mentions of a 'Sub-Editor' and 'Assistant Editor' in 1847 suggest that the staff complement increased somewhat – unless 'Assistant Editor' was just the reporter's job title.
4. Rainy, R. and J. Mackenzie, *Life of William Cunningham, D.D., Principal and Professor of Theology and Church History, New College, Edinburgh* (London, 1871), p. 138.
5. Sutherland and McKenzie Johnston, *Lydia*, pp. 172–73.
6. NLS MS.7516 f.188.
7. Guthrie, D. K. and C. J. Guthrie, *Autobiography of Thomas Guthrie, D. D. and memoir by his sons Revd David K. Guthrie and Charles J. Guthrie, M.A.* (London, 1874–75), vol. 2, pp. 222–23.
8. *Life*, vol. 2, pp. 231–32.
9. *Life*, vol. 2, p. 346.
10. Rainy and Mackenzie, *Cunningham*, p. 136.

JOHN SWANSON

Miller's boyhood friend, Free Church minister for the Small Isles, and master of the floating manse *Betsey*. Frontispiece to A. Cameron, 'Revd John Swanson', in Anon. [Greig, J.] (ed.) *Disruption worthies of the Highlands*, pp. 127–36 (Edinburgh, 1877).

(IMAGE © NATIONAL MUSEUMS SCOTLAND)

CHAPTER 10

His business was to fight[1]

MILLER, despite his earlier reservations, grew into his key role as Editor of the *Witness*: which was to be a controversialist and campaigner supporting the Evangelicals. Cunningham's biographers referred aptly to William Cobbett, the equally vigorous English journalist and author of *Rural Rides* (but, unlike Miller, a Radical politician), when recalling Miller's early days at the paper:

> A powerful combatant had entered the field, and all who dared to fight with him sunk under the weight of his blows. Rival editors he tomahawked and scalped. Despairing pamphleteers called him 'the sledge-hammer of the Non-Intrusion party'. His style was clear, strong, Cobbett-like English, rich with allusion and illustration, irresistible in mirth, and terrible in sarcasm.[2]

Guthrie alluded to the neighing warhorse of Job, chapter 39:

> His business was to fight, – and, like the war-horse that saith among the trumpets, Ha, ha, and smelleth the battle afar off, fighting was Miller's delight. On the eve of what was to prove a desperate conflict, I have seen him in such a high and happy state of eagerness and excitement, that he seemed to me like some Indian *brave*, painted, plumed, leaping into the arena with a shout of defiance, flashing a tomahawk in his hand, and wearing at his girdle a very fringe of scalps, plucked from the heads of enemies that had fallen beneath his stroke. He was a scientific as well as an ardent controversialist; not bringing forward, far less throwing away, his whole force on the first assault, but keeping up the interest of the controversy, and continuing to pound and crush his opponents by fresh matter in

every succeeding paper. When I used to discuss subjects with him, under the impression, perhaps, that he had said all he had got to say very powerful and very pertinent to the question, nothing was more common than his remarking, in nautical phrase, 'Oh, I have got some shot in the locker yet – ready for use if it is needed!'[3]

Lydia felt that her previously relaxed husband had changed, so their granddaughter reported many years later:

No one could be more utterly relentless, it seemed, in his castigation of an opponent, no one more keenly stinging and caustic in the manner of giving it. Yet the old gentle nature was always there deep down, and it was always the *view*, the *opinion* he really attacked, not the *man*. 'If you want me to crush any one,' he would say, half in jest and half in earnest, 'don't let me see him.'[4]

Indeed, if the *Life* is to be believed, his combativeness stemmed at least partly from prominent Evangelicals hounding him onto their mutual prey. Miller's journalism reads today as robust, even brutal, and on occasion even personally abusive. But Miller was, if anything, temperate by the standards of the time, for the Victorians relished vigorous exchanges.[5] The Evangelicals badly needed his sharp words if they were to campaign against patronage, which was still unresolved. Moreover, the Evangelicals, prevented by landed interests from properly reforming the parish system, had set up a number of the so-called *quoad sacra* churches to provide a spiritual service to otherwise neglected populations, over and above the normal parish kirks. These new churches' ministers and elders naturally tended to be Evangelical, leading to a further dispute over whether they could vote on Kirk business, especially at the Assembly.

In January 1843 the Government rejected the Evangelicals' 'Claim of Right' to the Church's spiritual independence from the law courts. Miller watched as Scotland's social and religious fabric was torn apart at that year's Assembly on 18 May:

The Moderator rose and addressed the House in a few impressive sentences. There had been an infringement, he said, on the constitution of the Church, – an infringement so great, that they could not constitute its General

Assembly without a violation of the union between Church and State, as now authoritatively defined and declared. He was therefore compelled, he added, to protest against proceeding further; and, unfolding a document which he held in his hand, he read, in a slow and emphatic manner, the protest of the Church. For the first few seconds, the extreme anxiety to hear defeated its object, – the universal hush, hush, occasioned considerably more noise than it allayed; but the momentary confusion was succeeded by the most unbroken silence; and the reader went on till the impressive close of the document, when he flung it down on the table of the House, and solemnly departed. He was followed, at a pace's distance, by Dr Chalmers; Dr Gordon and Dr Patrick McFarlan immediately succeeded; and then the numerous sitters on the thickly occupied benches behind filed after them, in a long unbroken line, which for several minutes together continued to thread the passage to the eastern door, till at length only a blank space remained.[6]

Thus did the Free Church of Scotland – a name which Miller was one of the first to use, and which he perhaps coined – leave the Established Church: or rather, as its members saw it, they took the Church with them. But some middle-of-the-road ministers who had been expected to 'come out' did not do so. One reason was that the Evangelical position (and Miller's) on patronage had shifted from allowing the congregation a simple veto of the patron's nominee, to complete abolition of patronage and direct election by all adult communicants – which, notably for the time, enfranchised women and working-class men. So, from 1195 ministers, just 454 went into the Free Church, and many elders and congregations: enough to damage the Establishment seriously, and remarkable given what the ministers and their families had to lose, but not enough to claim credibly that the Church of Scotland had taken itself out of the hands of State and laird.

The Disruption, as it was called, aroused considerable attention nationally and internationally, because of the stand on principle (especially by those ministers who had left particularly lucrative livings), and the great significance of the underlying issue, of whether a national Church could survive without State aid (though Miller was in favour of such aid, as a general principle). Miller and the *Witness* had been crucial in raising public support for the Evangelicals, and in return this public interest brought the newspaper's circulation to over three thousand in 1843, surpassing the *Scotsman*, but slipping back somewhat once the crisis had passed.

Much of Miller's reportage now concerned the Disruption's consequences, notably the need for financial support for the ministers' stipends, a new parallel network of churches, manses, schools, and missions at home and abroad, and a new theological college. To begin with, Free Kirk congregations simply met wherever they could. Miller wrote to Lydia from Cromarty in the summer of 1843:

> I do begrudge the Moderates our snug comfortable churches. I begrudge them my fathers' pew. It bears date 1741, and has held by the family, through times of poverty and depression, a sort of memorial of better days, when we could afford getting a pew in the front gallery. But yonder it lies, empty within an empty church, a place for spiders to spin undisturbed, while all who should be occupying it take their places on stools and forms in the factory close.[7]

Some lairds exploited the feudal elements of Scots tenurial law to deny Free Church congregations sites to build churches, manses and schools, particularly in the Highlands and Islands where the laird might own all the land for miles around. One solution was to worship outdoors, with the minister in the traditional wooden preaching tent (as it was called), and the congregation exposed to the elements. Even then, they might have to hold the service in the middle of the road or on the foreshore, in some areas the only land outside the laird's control. By no means all lairds were so determined, but such harassment was counterproductively offensive, recalling the Stuart persecutions of the conventicles of the seventeenth-century Covenanters. Such meetings were reminiscent also of Christ's preaching in Galilee, and the Israelites' wandering in the wilderness.

It so happened that Miller's old friend John Swanson was now Minister for the Small Isles of Inverness-shire, and when he 'came out' in the Disruption, the lairds would allow him no useable site for a house, let alone a church. So Swanson lodged his family on Skye, and tended his island parish from an old and dangerously leaky yacht, the *Betsey*. In reporting tours-cum-geological holidays in 1844 and 1845, Miller joined Swanson on the *Betsey*, whose cabin, replete with the tools of maritime and spiritual navigation, he

... found to be an apartment about twice the size of a common bed, and just lofty enough under the beams to permit a man of five feet eleven to stand erect in his nightcap. A large table, lashed to the floor, furnished with tiers of drawers of all sorts and sizes, and bearing a writing desk bound to it a-top, occupied the middle space, leaving just room enough for a person to pass between its edges and the narrow coffin-like beds in the sides, and space enough at its fore-end for two seats in front of the stove. A jealously-barred skylight opened above; and there depended from it this evening a close lanthorn-looking lamp, sufficiently valuable, no doubt, in foul weather, but dreary and dim on the occasions when all one really wished from it was light. The peculiar furniture of the place gave evidence to the mixed nature of my friend's employment. A well-thumbed chart of the Western Islands lay across an equally well-thumbed volume of Henry's 'Commentary'. There was a Polyglot and a spy-glass in one corner, and a copy of Calvin's 'Institutes', with the latest edition of 'The Coaster's Sailing Directions' in another; while in an adjoining state-room, nearly large enough to accommodate an arm-chair, if the chair could have but contrived to get into it, I caught a glimpse of my friend's printing-press and his case of types, canopied overhead by the blue ancient of the vessel, bearing in stately six-inch letters of white bunting, the legend, 'FREE CHURCH YACHT.'[8]

These trips (on one of which the *Betsey* almost foundered under Miller) led to some of Miller's finest reportage, later republished, with further travels in Scotland, in *Cruise of the Betsey, with Rambles of a Geologist*.

Miller was never a slavish propagandist for a party line. The *Witness* did usually support the Free Church and was broadly liberal or 'Whig' in politics. But when Chalmers, who strongly supported Miller, became less active in Kirk affairs, power passed increasingly to the Revd Robert Candlish and his allies who, during the mid-1840s, sought to bring the *Witness* under their direct control. This was anathema to Miller, who prized his independence and also disagreed with Candlish on many matters. In any case, in 1844 Miller himself took over the publisher's debt to the newspaper's original proprietors, thereby becoming co-proprietor with Fairly, with whom he had an excellent relationship. Although Miller now had a major financial commitment in the *Witness*, he was so concerned for his independence that he refused the bait of having the debt written off in return for being overseen by a Kirk nominee. Miller was also anxious to maintain the newspaper's commercial viability, especially now the Disruption was past. As he saw it, space was already under pressure from the heavy burden of Kirk business, but some Free Church ecclesiastics still pushed for the inclusion of even more long verbatim accounts of Kirk politics, often inserted at the last minute, to the disgust of the newspaper's staff.

Matters came to a head in 1846 when Candlish and his allies tried to merge the *Witness* with another Free Church newspaper, the *Scottish Guardian*, and place the resulting paper under their control. Miller reacted strongly with a trenchant memorandum to the proprietors in January 1847, calling upon Chalmers's support at the resulting meeting. Thereafter he was left to run the *Witness* as he thought best. He had kept his independence and self-respect, while his editorial perspective continued staunchly Presbyterian and specifically Free Kirker. But this was at the cost of further damaging his relations with Candlish and allies, who took control of the Free Church on Chalmers's death in May 1847.

Years later, Lydia (through Bayne) portrayed Miller as deeply wounded by the affair, and thereafter, together with the *Witness*, isolated from the Free Church and inactive in its debates. Lydia's assessment does not ring wholly true. Miller's social life in Edinburgh, which had never been as free and easy as at Cromarty, still included good friends in the Church, while he continued to comment on Church policy, such as on public education and on the training of ministers. Doubtless Miller was saddened when the Free Church went in directions which he thought mistaken. But without being an ordained minister he could do little to influence the Church's internal decision-making, especially after Candlish and his allies took power.

Some sense of Miller's exclusion can be gauged from *The Ten Years' Conflict* (1849), on the Disruption and its causes, by Candlish's ally the Revd Robert Buchanan. Here Miller's name was wiped from history. Miller gave the book a fair review, but others set the record straight. Guthrie, for one, stated flatly that Miller was

> … that mightiest champion of the truth, who did more to serve its cause than any dozen ecclesiastical leaders, and was beyond all doubt or controversy, with the exception of Dr Chalmers, by much the greatest man of all who took part in the 'Ten Years' Conflict'.[9]

Miller's role in the Disruption was one of his most significant impacts on Scotland's public life. He had indeed mobilised the public support and funding which enabled the Disruption. Whether he regretted it, at least as it turned out, is another matter: Chalmers certainly did. But there seemed no alternative at the time, and the Disruption was a victory for spiritual independence, principle, and the liberties of the subject. However, the rump Church of Scotland and the Free Church were now just two amongst several Reformed Churches, none so large as to claim a majority of Presbyterian churchgoers. The burden of church, manse and college building forced the Free Church away from its aspirations to be the true national Church and towards an introverted, essentially 'voluntarist', provision for just its members. As for the rump Established Church, the Disruption must have damaged its claim to speak for the whole community, accelerating the collapse of the old parish system and the creation of new secular services for education and social support. But we shouldn't blame the Disruption – or Miller – for the loss of the Godly Commonwealth, which was surely inevitable under the stress of social and economic change. And, far from suddenly disappearing, the underlying Presbyterian Christian ethos deeply influenced Scottish public services until well into the twentieth century.

NOTES TO CHAPTER 10

1. Guthrie and Guthrie, *Guthrie*, vol. 2, p. 2.
2. Rainy and Mackenzie, *Cunningham*, pp. 135–36.
3. Guthrie and Guthrie, *Guthrie*, vol. 2, pp. 2–3.
4. L. M. F. F. Miller, 'Mrs Hugh Miller's Journal', p. 516.
5. Cowan, R. M. W. *The newspaper in Scotland. A study of its first expansion 1815–1860* (Glasgow, 1947), pp. 244–46.
6. 'The Disruption', pp. 472–79 in *Headship*, pp. 476–77.
7. *Life*, vol. 2, p. 383.
8. Miller, H. *The cruise of the Betsey; or, A summer ramble among the fossiliferous deposits of the Hebrides* (Edinburgh, 1889), p. 11. 'Five feet eleven' is 5 feet 11 inches. The 'ancient' was an ensign, i.e. flag. The 'Polyglot' was a bible in multiple parallel texts in different languages, e.g. English and Greek.
9. Guthrie and Guthrie, *Guthrie*, vol. 2, p. 3.

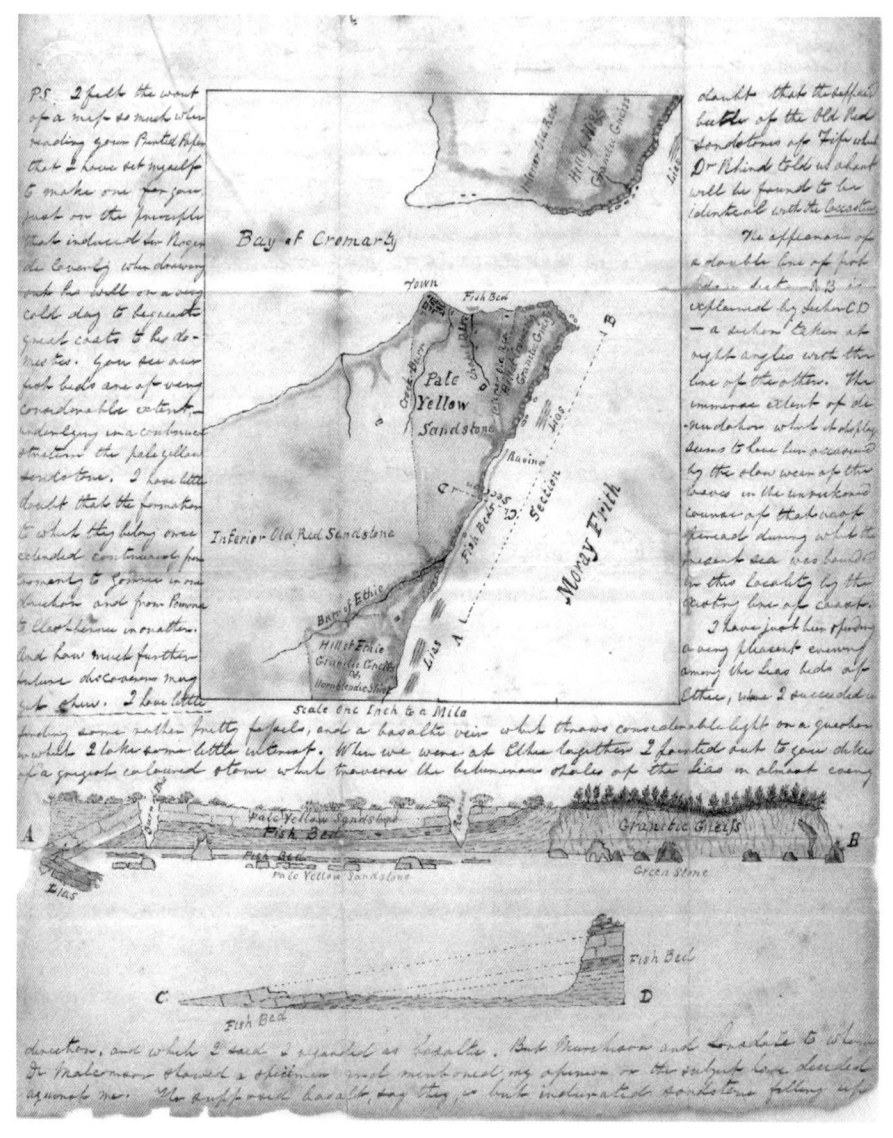

A LETTER TO PATRICK DUFF, 1839

Miller sent these sketch geological sections and map
of the Cromarty area on 27 January 1839.

(ELGIN MUSEUM, GEOLOGY LETTER G3/4)

CHAPTER 11

The truth I speak, impugn it whoso list

FOR the *Witness*'s masthead motto Miller adopted the famous retort of John Knox when reprimanded for forthrightly expressing his views to Mary Stuart, then Queen of Scots, at his trial for treason: 'I am in the place where I am demanded of conscience to speak the truth, and therefore the truth I speak, impugn it whoso list.'

Miller, too, took an independent position on all matters. Yet what was this truth that Miller was 'demanded to speak', his world-view? The anonymity of articles in the *Witness* poses real problems for establishing just what Miller wrote, and, for that matter, just which articles by others he actually agreed with. However, we often have evidence for his authorship in the form of an internal comment, or a positive attribution by his family, especially for those used in posthumous compilations.

As one would expect from its origin, the *Witness* gave extensive coverage to the Free Church, and to the other Christian denominations: the rump Established Church of Scotland and other Presbyterians, the Episcopalians and Anglicans (notably the Anglo-Catholics), and the Roman Catholics. Reformed Protestantism was naturally suspicious of Roman Catholicism (then, moreover, associated with repressive Continental regimes), and Miller was, unsurprisingly, critical of 'Popery' – but he was also critical of other denominations. In Cromarty he had, almost alone, refused to oppose the extension of the vote to Roman Catholics; and Bayne, himself a Free Churchman, asserted that Miller was no anti-Catholic bigot, although feeling obliged to give space to a certain popular contributor's 'perpetual doses of anti-Popish harangue. Nevertheless the bigots were not satisfied. The *Witness* was pronounced lukewarm in the Protestant cause. A new paper, called the *Rock*, – Miller called it a *trap* rock, was started. ...'[1] And an important history of the Scottish press suggested that the *Witness* declined somewhat in popularity in the 1850s partly because 'the cry of "No Popery" came much more stridently from other journals'.[2]

On wider issues, Miller tied himself to no political party. He described himself as a moderate progressive (and had considered 'Old Whig' as a name for the newspaper). He approved of the 1832 Reform Act which had given the parliamentary vote to people with a modicum of property (or renting it), such as Miller himself at Cromarty. But he was wary of universal suffrage:

> Let Chartism assert what it pleases on the one hand, and Toryism what it may on the other, the property-qualification of the Reform Bill is essentially a good one for such a country as Scotland. In our cities it no doubt extends the political franchise to a fluctuating class, ill-hafted in society, who possess it one year and want it another; but in our villages and smaller towns it hits very nearly the right medium for forming a premium on steady industry and character, and for securing that at least the mass of those who possess it should be sober-minded men, with a stake in the general welfare.[3]

Scotland at this time was changing from a largely agrarian country to one more centred on industry, and where even farming was increasingly mechanised. Miller observed the development of a rootless agricultural proletariat and, like many others, he found it hard to adjust to the rise of the new working classes in the urban areas and mining–industrial settlements. He perhaps claimed too much from his experiences with the Niddrie masons when expressing his scepticism about the trade unions. But he was not alone in being wary of the insurrectionary potential of contemporary society. His mentality was mainstream, apparently reflecting the contemporary perspective of thinking people rather than bigots. He had no love of lairds for their own sake, and in fact was balanced in despising unbridled aristocratic power as well as political radicalism. Both reeked of self-interest, which was not Miller's creed. Indeed, his personal lack of self-interest is apparent from the affair of his appointment to the bank, and (as will later be seen) his rejection of a lucrative Civil Service post. He did have his pride and his hope for literary immortality, but those do not demand trampling on others. When he gave 'advice to young working men desirous of bettering their circumstances', he said:

You will gain nothing by attending Chartist meetings. The fellows who speak nonsense with fluency at these assemblies, and deem their nonsense eloquence, are totally unable to help either you or themselves; or, if they do succeed in helping themselves, it will be all at your expense. Leave them to harangue unheeded, and set yourselves to occupy your leisure hours in making yourselves wiser men. Learn to make a right use of your eyes: the commonest things are worth looking at, – even stones and weeds, and the most familiar animals. Read good books, not forgetting the best of all: there is more true philosophy in the Bible than in every work of every sceptic … . You are jealous of the upper classes; and perhaps it is too true that, with some good, you have received much evil at their hands. It must be confessed they have hitherto been doing comparatively little for you, and a great deal for themselves. But upper and lower classes there must be so long as the world lasts; and there is only one way in which your jealousy of them can be well directed: do not let them get ahead of you in intelligence.[4]

Miller did not condemn property. He wrote approvingly of those who used wealth responsibly, such as the merchant William Forsyth whom he credited, not entirely accurately, with the eighteenth-century economic development of Cromarty. But some of his finest polemic, energised and informed by his experience of the hand-to-mouth life of the poor, was aimed at those lairds whom he saw as abusing their wealth and power. When Miller criticised Lowland farm-workers' housing, he recalled his lodgings at Niddrie:

But the cottage was an exceedingly humble one. It was one of a line on the way-side, inhabited chiefly by common labourers and farm-servants, – a cold uncomfortable hovel, consisting of only a single apartment, – by many degrees less a dwelling to our mind, and certainly less warm and snug, than the cottage of the west-coast Highlander. The tenant, our landlord, was an old farm-servant, who had been found guilty of declining health and vigour about a twelvemonth before, and had been discharged in consequence. He was permitted to retain his dwelling, on the express understanding that the proprietor was not to be burdened with repairs; and the thatch, which was giving way in several places, he had painfully laboured to patch against the weather by mud and turf gathered by the way-side. … With every heavy shower the rain found its way through, and the curtains of his two beds, otherwise so neatly kept, were stained by dark-coloured blotches. The earthen floor was damp and uneven; the walls, of undressed stone, had never been hard-cast; but, by dint of repeated whitewashings, the interstices had gradually filled up. They were now, however, all variegated by the stains from the roof.[5]

It was indecent to expect a large and growing family of both sexes, and perhaps a lodger, to live in such a one-roomed cottage: 'it will not do to speak of forty-pound impossibilities and twenty-pound inconveniences, when the morality of the country is thus at stake.' Far better for the lairds to do without a few fripperies: 'Pecuniary sacrifices *must* be made by the proprietary of the country, even should they have to part, in consequence, with one or two superfluous horses or a few supernumerary dogs.'[6] As for unmarried male farm workers, they subsisted in bothies of the kind which Miller had inhabited as a mason: 'We have seen more than the mere outsides of bothies, and know from experience, that though they may be fit dwellings for hogs and horses, they are not fit dwellings for immortal creatures, who begin in this world their education for eternity.'[7]

Miller's social critique had religious roots. He held a deep fear that such hopeless conditions drove people to despair and immorality, burdening society in the short term, but also threatening their souls for eternity:

Next to the family comes the dwelling. As dress is the clothing of the individual, so is the house the clothing of the family. It ought to be sufficient, – sufficient for all the purposes of family life, – for decency, for convenience, for warmth, for shelter, for washing and cooking, for retirement, and for the separation of the sexes. Here society has failed. It is idle to speak of sanitary reform, and almost idle to speak of moral reform, when we contemplate the dwellings of a large portion of the working population. We can no more expect propriety of conduct in the individual if we clothe him in rags, and keep him in rags, than we can expect propriety of conduct in a family that lives habitually in the wretched lodgments which disgrace our towns and cities.[8]

Like many at the time, Miller feared the growth of a debased underclass within the anonymity of urban life, amongst people uprooted from the social discipline and support of more settled communities. He was unhappy with the 1840s proposals for a workhouse-based Scottish Poor Law on the English model. This, he felt, would promote pauperism while taxing the working in order to support the idle. He much preferred the traditional parish-based charity directed to the genuinely 'deserving' under the supervision of ministers and elders, and which he believed to be morally superior.

Miller was no less ferocious in attacking lairds over their control of entry to land, as when, in 1847, a landowner closed off Glen Tilt, where in 1785 the pioneer geologist James Hutton found critical evidence for the origin of granite as a molten rock:

> The great sheep-farms were permitted, in the first instance, to swallow up the old agricultural holdings; and now the let shootings and game-parks are fast swallowing up the great sheep-farms. The ancient inhabitants were cleared off, in the first process, to make way for the sheep; and now the people of Scotland generally are to be shut out from these vast tracts, lest they should disturb the game. There is no exception to be made by cat-witted dukes and illiterate lords in favour of the man of letters, however elegant his tastes and pursuits; or the man of science, however profound his talents and acquirements, or however important the objects to which he is applying them. The Duke of Leeds has already shut up the Grampians, and the Duke of Atholl has *tabooed* Glen Tilt. ... If one proprietor shut up Glen Tilt, why may not a combination of proprietors shut up Perthshire? Or if one sporting tenant bar against us the Grampians, why, when the system of shooting-farms and game-parks has become completed, might not the sporting tenants united shut up against us the entire Highlands?[9]

Miller was outraged at the wasteful devotion of good land to hunting, and profoundly disturbed by the penal Game Laws and how ordinary people were forbidden to kill even those game animals which were destroying their crops:

> It must be miserable policy that balances against the lives of human creatures and the morals of thousands of our humbler people, the mere idle

amusements of a privileged class, comparatively few in number, and who have a great many other amusements ... .[10]

Hugh Miller was a proud Scot and a patriot but, as with many Scots then, this was in the context of a Scotland within the United Kingdom. He joined the General Committee of the National Association for the Vindication of Scottish Rights set up in 1853. This group aimed, not so much to dissolve the Union with England, but rather to seek proper implementation of the Treaty of Union, and the resolution of breaches such as unfair representation in Parliament, and, indeed, ecclesiastical patronage.[11]

Like many others then, Miller believed that Sunday should be preserved as the Sabbath, a weekly day of rest, worship, and reflection. Therefore he opposed Sunday travel, which broke the Sabbath for the travellers and also for the workers involved. A major controversy arose in 1846 when the Edinburgh and Glasgow Railway ran Sunday trains. Miller was so disturbed by this, not so much for its own sake but for the wider implications of such treatment of a cornerstone of moral conduct, that he composed a nightmarish vision of the future collapse of society if such behaviour was tolerated:

> The railway, I said, is keeping its Sabbaths. All around was solitary, as in the wastes of Skye. The long rectilinear mound seemed shaggy with gorse and thorn, that rose against the sides, and intertwisted their prickly branches atop. The sloe-thorn, and the furze, and the bramble choked up the rails. The fox rustled in the brake; and where his track had opened up a way through the fern, I could see the red and corroded bars stretching idly across. There was a viaduct beside me: the flawed and shattered masonry had exchanged its raw hues for a crust of lichens; one of the taller piers, undermined by the stream, had drawn two of the arches along with it, and lay adown the watercourse a shapeless mass of ruin, o'ermasted by flags and rushes. A huge ivy, that had taken root under a neighbouring pier, threw up its long pendulous shoots over the summit. I ascended to the top. Half-buried in furze and sloe-thorn, there rested on the rails what had once been a train of carriages; the engine ahead lay scattered in fragments, the effect of some disastrous explosion, and damp, and mould, and rottenness had done their work on the vehicles behind. ... There was an open space in front, where the shattered

fragments of the engine lay scattered; and here the rails had been torn up by violence, and there stretched across, breast-high, a rudely piled rampart of stone. A human skeleton lay atop, whitened by the winds; there was a broken pike beside it; and, stuck fast in the naked skull, which had rolled to the bottom of the rampart, the rusty fragment of a sword.[12]

W. M. Mackenzie, one of Miller's sharpest critics and a fellow Black Isler, expressed the view that 'On all subjects his mind had been made up before he left Cromarty …'.[13] Miller was, to be sure, almost forty when he moved to Edinburgh, his mindset overwhelmingly shaped by his experiences; but this is still an exaggeration. Nevertheless, Mackenzie's comment crucially emphasises Miller's Cromarty roots. Miller's vision seems to have looked to the past Scotland of small burghs in the countryside rather than to the new Scotland of mass industrial-isation. But he was hardly unique: his key Free Church allies, such as Chalmers, Guthrie, and Cunningham, also came from the small-town rural Lowlands, where the old Godly Commonwealth had come closest to the ideal of Knox and the Reformers. In contrast, this system had never been completely implemented in the Highlands, and had long broken down in the modern urban and industrial populations.

Miller's social critique had religious roots beyond his concern to help the poor to live a decent life. His Calvinist work ethic might be summed up as: work hard, spend little, save hard, avoid frivolity and waste, and take full advantage of education. Scottish Calvinism in particular tended to decouple social, economic and moral worth from one another: it was not the resources of social status, money, time, mental and physical ability, and so forth, which one owned (or inherited), which made one's moral worth, but what one did with them. Add the characteristic (if relative) egalitarianism of Scottish society, and one comes close to explaining Miller's attitude to life and work, such as his contempt for game-obsessed landowners.

Moreover, Miller's Calvinist theology was 'necessitarian', believing that this world was sinful and fallen, but could surely, and therefore must, be improved. (That some people would be unchangeably saved and others damned under the Calvinist doctrine of predestination did not affect the duty of Christians to act with charity: 'by their fruits shall ye know them', as the Sermon on the Mount put it.) Miller's lack of complacency deeply influenced his views on social matters, already informed by his actual experience of working-class life below the safety level which the middle classes normally never penetrated, and to where he would often return on his travels.

Miller lived in a different world from the class- and party-based politics of later decades. In that time and context, his position was coherent and consistent. Perhaps part of Miller's attraction to today's reader is that one is never quite sure whether he was looking forward to an earthly paradise, or harking back to the golden age of an older and more humane society, almost as if a Calvinist version of William Cobbett recalling the good old days of roast beef on every Englishman's table. Miller's views, if expressed rather more forcefully (as befitting a newspaper editor) than those of his middle-class contemporaries, were those which one might find in a particularly thoughtful and independent-minded burgher. His independent spirit and political stance placed him close to the centre ground of contemporary middle-class moral and social thinking, so it is not surprising that his newspaper and his books sold well. Hugh Miller, after all, was not far from the Victorian moral ideal.

NOTES TO CHAPTER 11

1. *Life*, vol. 2, p. 301. Miller puns on the old sense of 'trap' as a hard, fine-grained igneous rock common in Central Scotland.
2. Cowan, *Newspaper in Scotland*, p. 281.
3. *Cruise*, p. 167.
4. *Old Red Sandstone*, pp. 33–34.
5. 'The cottages of our hinds', pp. 187–99 in Miller, H. *Essays, historical and biographical, political and social, literary and scientific* (Edinburgh, 1890), p. 194.
6. 'The cottages of our hinds', pp. 187–99 in *Essays*, p. 199.
7. 'The bothy system', pp. 200–7 in *Essays*, p. 202.
8. 'Our working classes', pp. 143–49 in *Essays*, pp. 146–47.
9. 'Glen Tilt tabooed', pp. 112–21 in *Essays*, pp. 116–17.
10. 'The crime-making laws', pp. 243–53 in *Essays*, p. 252.
11. *Scotsman*, 19 November 1853, p. 4.
12. 'A vision of the railroad', pp. 327–36 in Miller, H. *Leading articles on various subjects* (Edinburgh, 1890), pp. 331–32.
13. Mackenzie, W. M. *Hugh Miller: a critical study* (London, 1905), p. 242. As regards Miller's formative years more generally, Dr David Alston's important study *My little town of Cromarty: the history of a northern Scottish town* (Edinburgh, 2006) was published as the 2007 edition of this book was in final proof, and I was therefore unable to draw upon it as it deserves. It throws considerable light on the social, political, religious and economic environment of Cromarty, on Miller's role as local commentator, historian, and folklorist, and on how his memory was preserved locally.

CHAPTER 12

The landscape was one without figures[1]

MILLER'S northern origins, shaggy appearance and woollen plaid often lead to the assumption that he was a Highlander – which, as noted, was not the case. However, in the 1840s the true Highlanders, the Gaelic population of the Highlands and Islands, became the subject of Miller's campaigning pen. In these debates, racial identity was a crucial issue, for the Highlanders were supposedly of the 'Celtic' race, distinct from 'Saxon' Lowlanders (and equally 'Saxon' English). We see something of Miller's attitudes to race when, in 1847, he met the little daughter of a Parsi convert to Christianity:

> …. sadly in want of a companion this evening, [she] was content, for lack of a better, to accept of me as a playfellow; and she showed me all her rich eastern dresses, and all her toys, and a very fine emerald, set in the oriental fashion, which, when she was in full costume, sparkled from her embroidered tiara. I found her exceedingly like little girls at home, save that she seemed more than ordinarily observant and intelligent, – a consequence, mayhap, of that early development, physical and mental, which characterizes her race. She submitted to me, too, when I had got very much into her confidence, a letter she had written to her papa from Strathpeffer, which was to be sent him by the next Indian mail. And as it may serve to show that the style of little girls whose fathers were fire-worshippers for three thousand years and more differs in no perceptible quality from the style of little girls whose fathers in considerably less than three thousand were Pagans, Papists and Protestants by turns, besides passing through the various intermediate forms of belief, I must, after pledging the reader to strict secrecy, submit it to his perusal: – 'My dearest Papa, – I hope you are quite well. I am visiting mamma at present at Strathpeffer. …'

> ... an hour spent with little Buchubai made it tell upon me more powerfully than ever before, – that there is in reality but one human nature on the face of the earth. ... We overlook, amid the diversities of form, colour and language, the specific identity of the human family. The Celt, for instance, wants, it is said, those powers of sustained application which so remarkably distinguish the Saxon; and so we agree on the expediency of getting rid of our poor Highlanders by expatriation as soon as possible, and of converting their country into sheep-walks and hunting-parks. It would be surely well to have philosophy enough to remember what, simply through the exercise of a wise faith, the Christian missionary never forgets, that the peculiarities of race are not specific and ineradicable, but mere induced habits and idiosyncracies engrafted on the stock of a common nature by accidents of circumstance or development ... .[2]

Miller conforms to the modern cliché of an intolerantly superior Victorian intelligentsia only in his implication of an uninformed state of mind. He blamed the degradation of such as the Irish peasantry, not on racial inferiority, but on the 'iron hand of oppression'.[3] He plainly rejected the views of those who asserted, and would assert, that (for example) white people were a species distinct from and superior to other humans, as in books such as *The Races of Man* (1850) by Robert Knox, the Edinburgh anatomist whom Burke and Hare supplied with murdered slum-dwellers.

Like the Irish, the Scottish Highlanders and Islanders were 'Celts' suffering from racist perceptions. Many people regarded their Gaelic language and way of life, including their recently extirpated clan-based tribal society, as primitive. But Miller was fascinated by the traditional customs he saw in the 1840s while accompanying John Swanson on the *Betsey*, such as transhumance, then almost extinct. Miller's account of a shieling on the isolated north-eastern coast of Eigg is now a precious survival:

> Rarely have I seen a more interesting spot, or one that, from its utter loneliness, so impressed the imagination. The shieling, a rude low-roofed erection of turf and stone, with a door in the centre some five feet in height or so, but with no window, rose on the grassy slope immediately in front of the vast continuous rampart [of a cliff above]. A slim pillar of smoke ascends from the roof, in the calm, faint and blue within the shadow of the precipice, but

it caught the sun-light in its ascent, and blushed, ere it melted into the ether, a ruddy brown. A streamlet came pouring from above in a long white thread, that maintained its continuity unbroken for at least two-thirds of the way; and then, untwisting into a shower of detached drops, that pattered loud and vehemently in a rocky recess, it again gathered itself up into a lively little stream, and, sweeping past the shieling, expanded in front into a circular pond, at which a few milch cows were leisurely slaking their thirst. … An island girl of eighteen, more than merely good-looking, though much embrowned by the sun, had come to the door to see who the unwonted visitors might be. … And as she set herself to prepare for us a rich bowl of mingled milk and cream, John and I entered the shieling. There was a turf fire at the one end, at which there sat two little girls, engaged in keeping up the blaze under a large pot, but sadly diverted from their work by our entrance; while the other end was occupied by a bed of dry straw, spread on the floor from wall to wall, and fenced off at the foot by a line of stones. The middle space was occupied by the utensils and produce of the dairy, – flat wooden vessels of milk, a butter-churn, and a tub half-filled with curd; while a few cheeses, soft from the press, lay on a shelf above. The little girls were but occasional visitors … the shieling had two other inmates, young women, like the one so hospitably engaged in our behalf, who were out at the milking … they lived here all alone for several months every year, when the pasturage was at its best, employed in making butter and cheese for their master, worthy Mr. McDonald of Keill.[4]

Miller attended a Free Church service in the Gaelic school on Eigg, a 'low dingy cottage of turf and stone':[5]

There were a few of the male parishioners keeping watch at the door, looking wistfully out through the fog and rain for their minister; and at his approach nearly twenty more came issuing from the place, – like carder bees from their nest of dried grass and moss, – to gather round him, and shake him by the hand. The islanders of Eigg are an active, middle-sized race, with well-developed heads, acute intellects, and singularly warm feelings. And on this occasion at least there could be no possibility of mistake respecting the feelings with which they regarded their minister.[6]

The reference to a bees' nest was precisely apt for a turf-roofed building. But its other connotations helped Miller's sympathetic portrayal of the Eigg people as hardworking, religious people who deserved fair treatment. Most newspapers had very different views of the Gaels, emphasising their supposedly ungrateful idleness, or the romantic bravery of the clan soldiers in the service of the Crown – or even both at once. It could certainly be convenient, particularly for those with interests in land, to portray Gaels as primitive barbarians hindering economic progress. During the eighteenth and nineteenth centuries, indeed, many Highlands and Islands landowners abrogated their tenants' traditional but unwritten rights, and took the communal farmland for sheep-farming. Many tenants were resettled on plots of land that were deliberately made too small to live off, ensuring a captive workforce for the lairds' industries such as fishing and gathering seaweed. Others emigrated to the Lowland cities and to foreign lands, some to reap success, others to find failure or an early death.

Those Highland Clearances, as they were called, were a long and complex process, but it is plain that some lairds made the effects of the area's real problems (including overpopulation, economic change and, soon, a potato famine) far worse than they might have been. Miller was in no doubt what he thought about the supposed right of property owners to do whatever they liked, even to the obliteration of whole communities. He was profoundly shocked when he visited the island of Rùm in 1844:

> All was solitary. We could see among the deserted fields the grass-grown foundations of cottages razed to the ground; but the valley, more desolate than that which we had left, had not even its single inhabited dwelling: it seemed as if man had done with it for ever. The island, eighteen years before, had been divested of its inhabitants, amounting at the time to rather more than four hundred souls, to make way for one sheep-farmer and eight thousand sheep. … there were fields on which the heath and moss of the surrounding moorlands were fast encroaching, that had borne many a successive harvest; and prostrate cottages, that had been the scenes of christenings, and bridals, and blythe new-year's days; – all seemed to bespeak the place a fitting habitation for man, in which not only the necessaries, but also a few of the luxuries of life, might be procured; but in the entire prospect not a man nor a man's dwelling could the eye command. The landscape was one

without figures. I do not much like extermination carried out so thoroughly and on system; – it seems bad policy; and I have not succeeded in thinking any the better of it though assured by the economists that there are more than people enough in Scotland still. … It did not seem as if the depopulation of Rum had tended much to anyone's advantage. The single sheep-farmer who had occupied the holdings of so many had been unfortunate in his speculations, and had left the island: the proprietor, his landlord, seemed to have been as little fortunate as the tenant, for the island itself was in the market; and a report went current at the time that it was on the eve of being purchased by some wealthy Englishman, who purposed converting it into a deer-forest. How strange a cycle! Uninhabited originally save by wild animals, it became at an early period a home of men, who, as the gray wall [of a deer trap] on the hill-side testified, derived, in part at least, their sustenance from the chase. … the island was to return to its original state, as a home of wild animals, where a few hunters from the mainland might enjoy the chase for a month or two every twelvemonth, but which could form no permanent place of human abode. Once more, a strange and surely most melancholy cycle![7]

Rùm had indeed returned to barbarism. And in articles reprinted as a pamphlet, *Sutherland as it was and is; or, How a Country may be Ruined* (1843), Miller attacked the powerful Duke and Duchess of Sutherland who owned almost all that county:

The county has not been depopulated – its population has been merely arranged after a new fashion. The late Duchess found it spread equally over the interior and the sea-coast, and in very comfortable circumstances; – she left it compressed into a wretched selvage of poverty and suffering, that fringes the county on its eastern and western shores.[8]

It is unclear why Miller did not sustain his campaign (beyond intermittent reports) into the 1850s when, if anything, the Highlands crisis worsened. This is a reason why, in one historian's view, he was 'far from a truly radical critic and a consistent missionary journalist for the Highland Gaels'.[9] But Miller had at least been an early and vocal defender. As a child, he had visited his aunt's family near

Lairg, in a cottage that was 'a low, long, dingy edifice of turf, four or five rooms in length, but only one in height, that, lying along a gentle acclivity, somewhat resembled at a distance a huge black snail creeping up the hill'.[10] The house may have had the cow-byre as the lowest room, and a living-room with a central open hearth, but Miller never forgot Cousin George, his role model, and the worthy and devout family therein. And when Miller, working as a mason at Gairloch, visited a peat-smoky Highland cottage, he was simply grateful to be given potatoes instead of the eternal bothy oatcake. Miller, that archetypal Lowlander, knew that Gaels were human beings too.

NOTES TO CHAPTER 12

1. *Cruise*, p. 135.
2. *Cruise*, pp. 354–55.
3. Miller, H. *The testimony of the rocks; or, Geology in its bearings on the two theologies, natural and revealed* (Edinburgh, 1890), p. 234.
4. *Cruise*, pp. 81–83.
5. *Cruise*, p. 90.
6. Ibid.
7. *Cruise*, pp. 133–36. A Highland 'deer-forest' did not necessarily have trees.
8. Pages 388–453 in *Leading Articles*, p. 392.
9. Fenyö, 'Views of the Highlanders', p. 326.
10. *Schools*, p. 95.

CHAPTER 13

The quiet enthusiasm of the true fossil-hunter[1]

THE move to Edinburgh changed Miller's geologising as it did everything else. He nevertheless found time to pursue his passion, although it had to be fitted into 'the mere leisure interstices of a somewhat busy life'.[2] Still, there were new sites in the Edinburgh district to explore on half-days, notably the Coal Measures and other Carboniferous rocks, and the raised beaches and other relatively recent deposits. And now that he had a long vacation each year, he could start more widely exploring Scotland's fossiliferous deposits.

Edinburgh also gave Miller far more scope for meetings with other geologists and access to institutional libraries, although we know relatively little about these aspects of his life. He was a member, and for a term President, of the Royal Physical Society of Edinburgh, an important local scientific society. And he also served as one of the secretaries for the Geological Section of the Edinburgh meeting of the British Association for the Advancement of Science in 1850.

The situation, however, was changing, the science becoming increasingly professionalised – although Miller apparently remained happy with his own path, well known for his geological writings and for his fossil collection. In 1852 the young Archibald Geikie plucked up courage to make himself known. Miller received him kindly, and soon invited him again, to Geikie's surprise and pride:

Witness Newspaper Office
Edinburgh 15th January 1852.

My dear Sir
I trust to be quite at leisure on the evening of Saturday and shall expect to see you at six o'clock to take a quiet cup of tea with me, and discuss a few geological facts. A return omnibus passes my house at nine in the evening for Edinburgh. Yours truly,
      Hugh Miller[3]

Miller suggested the island of Pabay as a suitable study site, encouraged Geikie to present his research at the Royal Physical Society, and recommended him to Sir Roderick Murchison as a good recruit for the Geological Survey then being extended to Scotland. Thus did Geikie, described by the historian David Oldroyd as 'from a certain perspective … Miller's most important geological discovery',[4] begin his long ascent within the British scientific establishment.

Miller developed the aim of writing a book on the geology of Scotland. Judging from the manuscripts eventually published as *Cruise of the Betsey*, it would have been more of a personal, even literary, tour of Scottish geology than a more formal scientific study: but that would have been the way Miller had chosen. His interests developed new focuses in the years after 1840, influencing the areas where he enjoyed seeking fossils and observing the field evidence in the open Scottish countryside. Miller was amongst the first to discover and publish on Old Red Sandstone plants – then amongst the oldest known plants. Miller also looked in particular at the Jurassic fossil plants of Scotland, following up his Eathie work with new finds further up the Moray Firth coast; his collection continues to engage scientists today. He still remained deeply interested in the Old Red Sandstone fishes with which he had originally made his mark, and sought them in areas other than Cromarty, such as in Caithness where he was shown around the local rocks by Robert Dick, Thurso's fossil-collecting baker.

One recurring interest, appropriate for someone born on a raised beach and (from 1854) living on another at Portobello, was 'that comparatively modern geologic agent, whatever its character, that crushed, as in a mill, the upper parts of the surface-rocks of the kingdom, and then overlaid them with their own debris and rubbish'.[5] Indeed, in the 1840s and 1850s the Ice Ages comprised, if one might so call it, a hot topic: not so much whether there had been an Ice Age, but its precise nature. There was much debate over the origin of features such as raised beaches, scratched rock surfaces, U-shaped valleys, and curious hillocks with tapering tails (so-called crags-and-tails), and the widespread superficial layers of clays, sand, and gravels. The older idea was of Scotland drowned under an icy sea, which had perhaps rolled in as a giant tsunami, and whose floating icebergs caught upon and ground over the submerged land. But, as reported to the Scottish public by Miller in early 1840, Professor Louis Agassiz argued, rather, that Scotland (and much of Europe) had been covered by a solid ice cap. Miller reserved judgement when contemplating examples of the 'mysterious scratchings and polishings now so generally connected with the glacial theory of Agassiz' in Kinloch Glen on Rùm in 1844:

> They testify, Agassiz would perhaps say, not regarding the existence of some local glacier that descended from the higher grounds into the valley, but respecting the existence of the great polar glacier. I felt, however, in this bleak and solitary hollow, with the grooved and polished platforms at my feet, stretching away amid the heath, like flat tombstones in a graveyard, that I had arrived at one geologic inscription to which I still wanted the key. … here was a story not clearly told. It summoned up doubtful, ever-shifting visions, – now of a vast ice continent, abutting on this far isle of the Hebrides from the Pole, and trampling heavily over it, – now of the wild rush of a turbid, mountain-high flood breaking in from the west, and hurling athwart the torn surface, rocks, and stones, and clay, – now of a dreary ocean rising high along the hills, and bearing onwards with its winds and currents, huge icebergs, that now brushed the mountain-sides, and now grated along the bottom of the submerged valleys. The inscription on the polished surfaces, with its careless mixture of groove and scratch, is an inscription of very various readings.[6]

Miller sought out evidence for climatic change such as Arctic species of molluscs from Fairlie on the Clyde, and fossil estuarine shells, wood and nuts from claypits near his (later) home at Portobello, attesting to a previous rise in sea level.

This emphasis on fossils was characteristic of his approach to geology. Miller found many fossils in natural exposures, such as along the coast. But the many small local quarries and mines then operating were also important sources of fossils, and there was indeed much to discover. Miller's collection derives part of its importance today from the fact that many of its fossils can no longer be collected, their sites now worked out or infilled. Today we can no longer seize the opportunities available to Miller at Clune quarry, a little south-west of Forres:

> … the fish-enveloping nodules, which are composed in this bed of a rich limestone, have been burnt, for a considerable number of years, for the purposes of the agriculturist and builder. There was a kiln smoking this evening beside the quarry; and a few labourers were engaged with shovel and pickaxe in cutting into the stratified clay of the unbroken ground, and

throwing up its spindle-shaped nodules on the bank, as materials for their next burning. … I asked one of the labourers whether he did not preserve some of the better specimens, in the hope of finding an occasional purchaser. Not now, he said: he used to preserve them in the days of Lady Cumming of Altyre; but since her ladyship's death, no one in the neighbourhood seemed to care for them, and strangers rarely came the way.[7]

Unlike many contemporaries' collections (some amassed partly by purchase), Miller's fossil collection shows little sign of collecting for show. There is no obvious bias towards complete specimens primarily for display (although, as will be seen, he did have an eye for the beauty of fossils). It is very much the collection of a serious collector focussed on his interests: it has much fragmentary and aesthetically unimpressive but scientifically valuable material. After all, Miller knew from the Cromarty fishes how useful even fragmentary remains could be. He looked at his fossil specimens at all levels of detail, owning a portable 'botanist's microscope' – probably simply a high quality lens on a stand – and having slices cut from some of his fossils, ground down and polished, and mounted onto microscope slides for further study. Miller did not seek kudos by giving away large numbers of specimens to institutions (although he was pleased to see some of his fossils in the British Museum [now Natural History Museum, London]). Little seems to have been purchased, except perhaps to fill gaps, or when a special reason arose: say, to secure a deceased collector's Burdiehouse fishes or to pay a Girvan collector who guided him around the local sites.[8] Miller also received some fossils from abroad, presumably as presents, from Free Church missionary and evangelical workers, and other exiles in the Scottish Diaspora.

Miller's specimens, today preserved in National Museums Scotland, represent an immense amount of hard work, of exploration, searching and discovery, and subsequent preparation at the workbench. These fossils comprise perhaps his most significant formal contribution to science. Many specimens were used by Miller, his contemporaries, and later workers, in their publications, and more yet will be. The most important are, as I have already noted, the type specimens on which the names of genera and species are defined. But any specimen discussed in a scientific paper becomes valuable: just as an experimentalist's work is valued only if the experiment can be replicated by other workers, it is the fact that the specimen is in a museum which enables a palaeontologist to replicate a colleague's observations. Nor should one ignore any of Miller's field observations on the sites themselves; they might have received more attention were it not for a shift away from palaeontology in the second half of the 19th century – at any rate, his observations on the Isle of Eigg, for example, were not fully appreciated or followed up for more than a century.

At Edinburgh, therefore, Miller's knowledge of geology continued to develop and build upon his exposure to the wider world of geology triggered by Malcolmson. However, what set him apart from other geologists was not so much the quality of his science – or of his fine fossil collection, important as it was and remains – but the same literary activity that, ironically enough, limited his geologizing. Miller's engagement with geology was always twofold: his publications being the public face, while the more private aspect comprised his direct contact with geology – wandering the landscapes, observing and thinking about the strata, and finding fossils to take home and study. But, of course, these two threads were interdependent. As well as their raw matter, Miller's writings drew enormous power and authority, leavened by his attractive autobiographical perspective, from his personal fieldwork and his collections. This synergy enabled Miller to turn his geological wanderings into writings which, quite apart from their real economic value, might help achieve some of the immortality which he felt he had forsaken when he joined the newspaper. It is now appropriate to turn to those writings, which conferred that immortality upon Miller's specimens as much as upon the man himself.

NOTES TO CHAPTER 13

1. *Cruise*, p. 193.
2. *Cruise*, p. 476.
3. A. Geikie, *A Long Life's Work* (London, 1924), pp. 24–25.
4. Oldroyd, 'The geologist from Cromarty', p. 109.
5. *Cruise*, p. 233.
6. *Cruise*, pp. 123–24.
7. *Cruise*, pp. 191–92; Lady Gordon Cumming was a noted collector of fossil fishes.
8. G. Lyon, 'Memoir of the Society', *Transactions of the Edinburgh Geological Society*, **1**, 2–3 (1866); H. Miller, 'On the ancient Grauwacke rocks of Scotland', pp. 297–324 in *Old Red Sandstone*, p. 311; M. McCance, 'Hugh Miller, 1802–56, geologist and writer: his links with nineteeth-century Girvan', *Ayrshire History* **23**, 20–23 (2002).

THE OLD CHAPEL BURN EMERGING FROM THE DEN WHICH
IT HAS ERODED INTO THE ANCIENT SEA-CLIFF NEAR CROMARTY

Miller hunted for fossils everywhere around Cromarty, such as the rocks exposed in the bottom of this miniature ravine. The mossy stone dome, middle right, is that of the Coalheugh Well, a spring of water under pressure created by geologically ignorant miners hunting unsuccessfully for coal.

(MICHAEL A. TAYLOR)

CHAPTER 14

He clothed the dry bones of science[1]

ARCHIBALD Geikie remembered Hugh Miller, a personal hero of his, as clothing 'the dry bones of science with living flesh and blood'.[2] So it may seem odd that Miller is absent from the index of Geikie's classic history, *The Founders of Geology*. But, of course, Geikie was there chronicling those, such as James Hutton and William Smith, who made fundamental contributions to the science, and who changed the very way in which scientists thought. Miller's contribution to the science was rather different. He also fundamentally affected minds – and that included the minds of individuals such as Geikie, but before – and while – they became scientists. For, in many respects, Miller was the greatest geological evangelist of his time. I now assess his literary geology, as it developed after the success of *The Old Red Sandstone*.[3]

Miller's position as editor limited his geologizing but gave him ample opportunity – and incentive in the newspaper's insatiable appetite for copy – to write about his favourite science. This work was often serialised in his newspaper prior to publication in book form. That is not to say that the *Witness* was ever a geological journal – a reader was always far more likely to find the Free Kirk than a fossil; but its editor's fossilising proclivities would have been obvious from occasional essays and reviews. Miller also produced other writings which went straight into book form. He thus created a very special corpus of writings to which Geikie, no mean writer himself, paid a remarkable tribute:

> I do not think that the debt which geology owes to him for these services, in deepening the popular estimation of the science, and in increasing the number of its devotees, has ever been sufficiently acknowledged. During his lifetime, and for some years afterwards, Hugh Miller was looked upon by the general body of his countrymen as the leading geologist of his day. And this exaggerated but very natural estimate spread perhaps even more exten-

sively in the United States. His books were to be found in the remotest log-hut of the Far West, and on both sides of the Atlantic ideas of the nature and scope of geology were largely drawn from them.[4]

As Geikie implied, the natural limitations of such writing made it difficult to sort out Miller's own observation and analysis from those of others which he was simply reporting (though Miller might, in any case, have checked their work by his own observation and logic). But Miller's writing differed from academic treatises in far more than its lack of a detailed apparatus of references. He was addressing much wider audiences and dealing with much broader themes, and he developed a literary repertoire to suit his needs.[5]

To modern eyes Miller's books, such as the *The Old Red Sandstone*, seem slightly odd in their infusion of religion even in the geological sections, especially as, by 1841, such discussion was indeed obsolete in scientific literature. There is, for instance, that interpretation of the Day of Judgement as the last great geological 'catastrophe' (if seemingly presented more as the simple end of geological – and human – history, than as a roaring Apocalypse). This might lead the modern reader to assume something resembling modern creationism. But such a modern perspective is misleading, as it so often is with Miller. During the middle nineteenth century, many men – and many also of the then rather less overtly visible women – of science were Christians who believed in the truths of the Bible. As we have already seen, they saw no great contradiction with geology. Writers of geological treatises commonly included religion in their prefaces and conclusions, and Miller was only slightly unusual in also adding it to his geological discussions. He was, in fact, no religious extremist. He did not put that much divinity into his geological works, except when he dealt with the special theme of the relationship between the truths of science and those of religion (which was not the simple adversarial one that is often assumed, as will be seen in the later discussion of Miller and early ideas on evolution). In all this Miller was simply being practical. He was writing not for a formal scientific audience, but for a general – and largely Christian – public who wanted to know about the latest scientific finds and ideas and what these meant for their understanding of life and the universe. Indeed, his geological writing for the *Witness* would not have been aimed solely at that newspaper's Presbyterian readers, given that most of it was plainly intended for future publication in book form, in England as much as in Scotland.

Miller often started with the simple and familiar, before going into the complex and exotic. This had real communicative benefits, and gave him plenty of scope for his favourite device of the illustrative simile or metaphor – particularly beneficial in books that were usually only minimally provided with actual pictures. Certainly Miller had an eye for the picturesque. When he wrote of the 'vast denudation' of the red sandstone of Wester Ross, cut by erosion into the isolated hills of Suilven, Cùl Mor and Cùl Beg, resting on the grey gneiss below, he recalled watching them, with a painter's eye as much as a geologist's,

> ... at sunset in the finer summer evenings, when the clear light threw the shadows of their gigantic cone-like forms far over the lower tract, and lighted up the lines of their horizontal strata, till they showed like courses of masonry in a pyramid. They seem at such times as if coloured by the geologist, to distinguish them from the surrounding tract, and from the base on which they rest as on a common pedestal. The prevailing gneiss of the district reflects a cold bluish hue, here and there speckled with white, where the withered and lichened crags of intermingled quartz-rock jut out on the hillsides from among the heath. The three huge pyramids, on the contrary, from the deep red of the stone, seem flaming in purple.[6]

And there is no better outline of geology's economic importance than Miller's account of an old attempt to mine coal in the Old Red Sandstone at Cromarty, futilely so as the Old Red lies *below* the Coal Measures:

> [The laird] accordingly brought miners from the south, and set them to bore for coal in the gorge of the ravine. Though there was probably a register kept of the various strata through which they passed, it must have long since been lost; but, from my acquaintance with this portion of the formation ... I think I could pretty nearly restore it. They would first have had to pass for about thirty feet through the stratified clays and shales of the ichthyolite bed ... in time, if they continued to urge their fruitless labours, they would arrive at the primary rock, with its belts of granite, and its veins and huge masses of hornblende. In short, there might be some possibility of their penetrating to the central fire, but none whatever of their ever reaching a vein of coal.[7]

Miller here soon moved from particular to general, from concrete to abstract, in a passage which, as so often with him, penetrated to the heart of the matter – although his final destination was never one of pure abstraction:

> There is no science whose value can be adequately estimated by economists and utilitarians of the lower order. Its true quantities cannot be represented by arithmetical figures or monetary tables; for its effects on mind must be as surely taken into account as its operations on matter, and what it has accomplished for the human intellect, as certainly as what it has done for the comforts of society or the interests of commerce. Who can attach a marketable value to the discoveries of Newton? ... And geology in a peculiar manner supplies to the intellect an exercise of this ennobling character. But it has also its cash value. The time and money squandered in Great Britain alone in searching for coal in districts where the well-informed geologist could have at once pronounced the search hopeless, would much more than cover the expense at which geological research has been prosecuted throughout the world.[8]

Miller sought to explain, not merely physical relationships, but logical and temporal ones. He clearly expounded (but did not originate) the fundamental geological concept of actualism, which Archibald Geikie later reprised, famously amongst geologists, as 'The present is the key to the past':[9] given that geological processes today are the same as in the past, then the phenomena in the rocks can be explained by observation of present-day processes (at least, 'in between' the rare catastrophes, which, as we have seen, were also invoked to explain what appeared to be unusually intense phenomena).

Thus Miller invited 'the reader to accompany us' in a hop over a muddy hole in an Edinburgh park in 1842, an imaginative – but always evidentially controlled – step into the geologically recent past, and then a leap beyond into the deep time of the Carboniferous:

> We pass onwards, and reach a different subsoil. The leaden-coloured tint has given place to one of deep black, with here and there a mingling band of brown. We have arrived at one of the beds of lacustrine peat, in which the Meadows abound. The Irish labourers employed in the work of draining,

reminded apparently of their green bogs, and their weeks of periodical summer labour spent in restocking their turf-stacks, exhausted during the chills of winter, have been cutting the masses in the peat form, and placing them end-long, in pairs, all along the margin of the ditch. ... When the Borough Loch began to shallow, it must have formed a swampy patch, whose surface, level with that of the water, would seem in the drier seasons a low island, and, after prolonged rains, would return to its original condition of a dark tangled shoal. ... In this pale marly mass we find the remains of a flattened reed. Two of the joints are still discernible, with the minute striae that run between them; and the entire organism, both from its well-marked form and its deep black colour, contrasted with the pale tint of the surrounding mass, serves strongly to remind us of a calamite of the Coal Measures, enclosed in a smoke-coloured limestone. We have disinterred fossils in Burdiehouse – slender calamites – carbonised in the pale, splintery stone, that could scarce be distinguished from that recent reed in its matrix of marl.[10]

Miller made engaging diversions into folklore, history, and social comment, as in a mine at Dryden, south of Edinburgh, in the 1840s; for him, geology was no abstract concern but something real in the sense that it formed a landscape inhabited by people:

The time chosen for our exploratory descent is that when the workmen are absent at their mid-day meal; nor would it be very convenient to thread a passage so narrow and long, so broken by descents, and perplexed by turnings, when the bearers are passing out and in, either laden with coal, or carrying on their shoulders the bulky basket. Here the miners come, however, each furnished with his lamp. They squat themselves down in a reclining posture, which only habit could have rendered bearable; and, striking out the projecting masses of coal, roll them over into the gallery. And now come the bearers with their baskets, to carry up to the surface the coal thus disengaged. But what work for women? Each bears a lamp fastened to her head, to light the long upward ascent; and, laden with more than a hundredweight of coal, and bent forward at nearly a right angle, to avoid coming into contact with the low roof, they ascend slowly along the flights of steps, and through the narrow galleries, and, lastly, up the long stair of the shaft; and when they have reached the surface, they unload at the coal-heap, and return. And such is the employment of females for twelve, and sometimes fifteen hours together.[11]

Thanks to Miller's judiciously autobiographical tone and the spoken cadences of his writing, readers always felt as if they were in his personal company: perhaps touring a museum, or strolling along the shore, whether of the Cromarty Firth or a long gone ocean, or maybe as a traveller through time astounded by the Carboniferous swampland, like an intrepid Victorian naturalist seeing the rainforest from his ship. Miller rendered the static fossils into a dynamically imagined world:

> The water is fast shallowing. Yonder passes a broken branch, with the leaves still unwithered; and there floats a tuft of fern. Land, from the mast-head! land! land! – a low shore thickly covered with vegetation. Huge trees of wonderful form stand out far into the water. There seems no intervening beach. A thick hedge of reeds, tall as the masts of pinnaces, runs along the deeper bays, like water-flags at the edge of a lake. A river of vast volume comes rolling from the interior, darkening the water for leagues with its slime and mud … .[12]

The Victorians, of course, appreciated Nature at all levels, from landscape, to fossil, to the minutest detail of a single fish scale. Miller happily guided the reader around them all – quarry, museum, and microscope slide. The specimens were often from his collection and individually recognisable today. Even when escorting his readers into the depths of time, Miller always returned to the actual cliffs, rocks and specimens that made up the solid evidence for the past. For him these physical signs were like a library to be read and imagined, and also to be valued for their power in this regard. Here he is making eloquent strange indentations on sandstone: footprints left by a reptile walking

> … adown a moist sandy slope, furrowed by ripple-markings, apparently to a watering-place. He travelled leisurely, as became a reptile of consequence, set down his full weight each step he took, and left a deep-marked track in

double line behind him. And yet, were his nerves less strong, he *might* have bestirred himself; for the southern heavens were dark with tempest at the time, and a thunderous-like shower, scarce a mile away, threatened to wet him to the skin. On it came; and the large round drops, driven aslant by a gale from the south, struck into the sand like small shot, at an angle of sixty. How the traveller fared on the occasion has not transpired; but clear and palpable it is that he must have been a firm fellow, and that the heavy globular drops made a much less marked impression on the sand consolidated by his tread than when they fell elsewhere on the incoherent surface around him.[13]

The Victorians also adored panoramas, dioramas, and similar scenic illusions, and Miller's intensely visual writing took full advantage of those cinema-like effects. Like a diorama impresario, Miller showed scene after scene in dizzy sequence, each new scene changing while the stage is obscured in theatrical fog and darkness: 'The geologic diorama abounds in strange contrasts.'[14] Here is Miller's presentation of the Bass Rock, an island in the Firth of Forth. Today it is a gannet colony; in the seventeenth century it was a prison for Covenanters; but at the dim and distant start of geological history –

The curtain rises, and there spreads out a wide sea .... It is the ocean of our Scotch Grauwacke that rolls beneath and around us; but regarding its inhabitants, – so exceedingly numerous and well-defined in the contemporary seas of what is now England, – we can do little more than guess.[15]

And after a sequence of scene changes, he fired off a special effect to wake up the audience:

The land of the Coal Measures has ... disappeared; and a shoreless but shallow ocean, much vexed by currents, and often lashed by tempest, spreads out around, as during the earlier periods. But there are more deeply-seated heavings that proceed from the centre of the immediate area over which we stand, than ever yet owed their origin to storm or tide. Ever and anon waves of dizzy altitude roll outwards towards the horizon, as if raised by the fall of some such vast pebble as the blind Cyclops sent whizzing through the air after the galley of Ulysses, when

'The whole sea shook, and refluent beat the shore.'

We may hear, too, deep from the abyss, the growlings, as of a subterranean thunder, loud enough to drown the nearer sounds of both wave and current. And now, as the huge kraken lifts its enormous back over the waves, the solid strata beneath rise from the bottom in a flat dome, crusted with shells and corals, and dark with algae. The billows roll back, – the bared strata heave, and crack, and sever, – a dense smouldering vapour issues from the opening rents and fissures; and now the stony pavement is torn abruptly asunder, like some mildewed curtain seized rudely by the hand, – a broad sheet of flame mounts sudden as lightning through the opening, a thousand fathoms into the sky, –

'Infuriate molten rocks and flaming globes,
Mount high above the clouds,' –

and the volcano is begun. Meanwhile, the whole region around, far as the eye can reach, heaves wildly in the throes of Plutonic convulsion. Above many a rising shallow, the sea boils and roars ... .

And after yet more scenes, Miller came to the end:

My history speeds on to its conclusion. We dimly descry, amid fog and darkness, yet one scene more. There has been a change in the atmosphere; and the roar of flame and the hollow voice of earthquake are succeeded by the howling of wintry tempests and the crash of icebergs.[16]

Miller was not the only writer splendidly interpreting geology for the public. William Buckland (Oxford don and describer of the dinosaur *Megalosaurus*) and Gideon Mantell (Sussex surgeon with *Iguanodon* to his credit) were amongst an extensive cast of discoverers of new knowledge who also courted public attention for their finds and interpretations. The public was keen for news of the unbelievably extraordinary ancient worlds being unearthed, so to speak, in their back gardens. The science was indeed so popular that major scientific books, such as Charles Lyell's *Principles of Geology* (1830–33), were written with a mind to the general commercial market. But Miller's contemporaries perceived his special literary qualities; Buckland himself said that he would give his 'left hand to possess such powers of description'.[17] As Miller himself argued, a 'truly great poet' could

> ... find sermons in stones, and more of the suggestive and the sublime in a few broken scaurs of clay, a few fragmentary shells, and a few green reaches of the old coast line, than versifiers of the ordinary calibre in their once fresh gems and flowers, – in sublime ocean, the broad earth, or the blue firmament and all its stars.[18]

At times, Miller seems to the modern reader to make few concessions, as when he discusses rather technical details of his fishes and uses the formal scientific names without apology. This reflected the Victorian love of detail, and the Victorian reader's stamina. But it also showed Miller's insistence that the public deserved the most rigorous treatment, whether of Sunday sermons from competent ministers, or of the key issues of the day in their everyday newspapers. Rather than something diluted, perhaps distorted, and then regurgitated, Miller sought to give the real thing, for he had grown up on it:

> Those intolerable nuisances the useful-knowledge books had not yet arisen, like tenebrious stars, on the educational horizon, to darken the world, and shed their blighting influence on the opening intellect of the 'youthhood'. ...[19]

So it is ironic that today Miller's writing risks being dismissed as popularisation of science, a genre today often derided as neither proper science nor proper literature. In their more enlightened time, the Victorians did not divide science from culture, and had no problems in accepting Miller as a major literary figure, like Charles Dickens and Thomas Carlyle. He was not so much a literary geologist as an astonishingly rounded figure who wrote about his world in which geology happened to be an integral part. Miller's point, perhaps, was not so much to popularise geology to the wider world, as to change the science itself by making it central to his writings on all subjects and thereby write geology into everything. That was no small mission, yet it was deeply appropriate, for science was raising fundamental questions with huge implications for humanity's view of its own place in the world.

NOTES TO CHAPTER 14

1. Anon., *Centenary*, p. 61.
2. Ibid.
3. Miller's literary geology is discussed at much greater length in Ralph O'Connor's *The Earth on show: fossils and the poetics of popular science, 1802–56*. I am grateful to Dr O'Connor for a copy of the MS. I received this as the 2007 edition of this book was in final revision, so I was unable to draw upon this work as fully as it deserves.
4. W. K. Leask, *Hugh Miller* (Edinburgh, 1896), pp. 148–49.
5. O'Connor, *The Earth on show*.
6. *Old Red Sandstone*, p. 53.
7. *Old Red Sandstone*, p. 194.
8. *Old Red Sandstone*, p. 196.
9. For instance, A. Geikie, *Landscape in History and Other Essays* (London, 1905), p. 5; Oldroyd, 'The geologist from Cromarty'.
10. 'The Meadows', pp. 134–47 in Miller, H. *Edinburgh and its neighbourhood, geological and historical: with The geology of the Bass Rock* (Edinburgh, 1891) pp. 138, 139–42.
11. 'Geological features', pp. 1–90 in *Edinburgh*, pp. 73–74.
12. *Old Red Sandstone*, pp. 267–68.
13. Miller, H. *First impressions of England and its people* (Edinburgh, 1889) p. 191.
14. *Sketch-book*, p. 117.
15. 'Geology of the Bass', pp. 245–344 in *Edinburgh*, p. 326.
16. 'Geology of the Bass', pp. 245–344 in *Edinburgh*, pp. 338–39, 343; it quotes from Pope's 'Odyssey' and John Philips's 'Cyder'. The 'kraken' is, obviously, a purely metaphorical sea-monster.
17. Anon., *Centenary*, p. 57.
18. *Sketch-book*, p. 80.
19. *Schools*, pp. 28–29.

CHAPTER 15

Exceedingly plausible and consummately dangerous[1]

EVEN before Hugh Miller was born, certain answers had been suggested to questions about the origins of the world, life and humanity: answers which seemed subversive of religion, morality and society to many, especially those who benefited from the status quo. From the Bible, some scholars concluded that the world and its living things, human beings last of all, were divinely created in six days at a time around four thousand years before the birth of Christ. But during the eighteenth century, the Scot James Hutton found evidence that the world was far older, unmeasurably so; while other thinkers – such as the author of *Telliamed* which the young Miller read – suggested that living things had evolved from one another, instead of being divinely created as fixed species. The resulting, politically charged, debate was never a simple one of science versus religion, or of progressive scientist versus religious reactionary, or, for that matter, of social moralist versus revolutionary atheist – for, by the 1840s, pre-Darwinian evolution had strong political overtones, partly because it had some of its roots in Revolutionary France. To be sure, anything which denied God a role in creation was indirectly an attack on the status quo shored up by the Established Churches. But what complicates the story is that, as well as some of the bitterest opposition, some of the strongest support for science came from the Churches. Of that support, Hugh Miller is an outstanding example.

God was always there, so to speak, in everything Miller wrote. But He had a special place in geology, which was part of 'natural theology', evidence in the natural world for a Divine Creator, which complemented 'revealed theology' from direct Divine revelation such as the Bible. Thus, when Miller explored the detailed anatomy of fossil fishes, he saw the Divine Designer as a Scots craftsman:

> The slater fastens on his slates with nails driven into the wood: the tiler secures his tiles by means of a raised bar on the under side of each, that locks

into a corresponding bar of deal in the framework of the roof. Now in some of the scales I found the art of the tiler anticipated: there were bars raised on their inner sides, to lay hold of the skin beneath; while in others it was the art of the slater that had been anticipated, – the scales had been slates fastened down by long nails driven in slantwise, which were, however, mere prolongations of the scale itself. Great truths may be repeated until they become truisms, and we fail to note what they in reality convey. The great truth that all knowledge dwelt without beginning in the adorable Creator must, I am afraid, have been thus common-placed in my mind; for at first it struck me as wonderful that the humble arts of the tiler and slater should have existed in perfection in the times of the Old Red Sandstone.[2]

The abstract beauty of fossils also stemmed from their divine creation, as with a coral's mathematical elegance heralding a fashionable fabric print:

The beautifully-arranged lines which so smit the dames of England, that each had to provide herself with a gown of the fabric which they adorned, had been stamped amid the rocks *eons* of ages before.[3]

Miller, moreover, recommended geology as an improving recreation exploring the works of the Lord. It exercised the body with hard open-air work, and the mind with the scientific study of the finds. But none of this was unusual, except perhaps in the deftness of Miller's prose, and his palaeontological aesthetics. His contemporaries, in both science and in the Churches, routinely praised natural history as an improving occupation. Indeed, natural theology was taken seriously at the Free Church College established in Edinburgh in 1843. Although the College trained students only for the ministry, it soon had its own museum of natural science, and, for its Professor of Natural Science, one of the finest natural scientists in Scotland, the Revd Professor John Fleming. One reason was to teach the trainees how to cope with evolutionists – for, of course, the spectre of evolution had never gone away.

Far from it, for there was published in 1844 *Vestiges of the Natural History of Creation*, written anonymously by Miller's fellow Edinburgh journalist and amateur geologist Robert Chambers. Aimed at a progressive middle-class Victorian audience, it notoriously argued for the evolution of the universe and life, pre-programmed by the Divine Creator to culminate in humanity. The book caused sufficient outrage that suspicion of Chambers's authorship prevented him from becoming Lord Provost of Edinburgh. Miller delayed a full-scale response, lest he attract even more attention to the book. But when a cheap edition of *Vestiges* was published for mass sale, Miller attacked head-on in *Foot-prints of the Creator* (1849). This was a new departure for Miller, in that *Foot-prints* was far better illustrated than his earlier books. It necessarily focussed on fairly detailed refutations of *Vestiges*'s assertions (and, if only for this reason, is perhaps the least readable of Miller's books). It used Miller's observations, and often his fossils, combined with discussions of the religious and social implications of evolutionism: to Miller, 'a form of error at once exceedingly plausible and consummately dangerous, and which is telling so widely on society, that one can scarce travel by railway or in a steamboat, or encounter a group of intelligent mechanics, without finding decided trace of its ravages'.[4]

Miller pursued similar arguments in his other book on geology and religion, *Testimony of the Rocks* (1857), which contains some of his best writing, less constrained by the need to argue with others. *Foot-prints* and *Testimony* were important in their day. They show that Miller was never torn between science and religion, contrary to what is sometimes asserted. Rather, Miller held a middle course between biblical literalists and those geologists and materialists who saw no role for God. One of his most scathing essays targeted those religious reactionaries whom Miller accused (not always entirely fairly) of denying the validity of scientific evidence. It was absurd to compromise between science and religion, he argued, for both were unquestionably part of the same Divine truth:

> Between the Word and the Works of God there can be no actual discrepancies .... The geologist, as certainly as the theologian, has a province exclusively his own; and were the theologian ever to remember that the Scriptures could not possibly have been given to us as revelations of scientific truth, seeing that a single scientific truth they never yet revealed, and the geologist that it must be in vain to seek in science those truths which lead to salvation, seeing that in science these truths were never yet found, there would be little danger even of difference among them, and none of collision.[5]

Miller did not read Genesis literally, for he, like many churchmen (Thomas

Chalmers included), accepted the evidence for the great age of the Earth. He argued, ingeniously, that Genesis was no mere origin myth, but an actual, divinely inspired, vision of the successive stages of geological time. As for Noah's supposedly global Flood, it was a regional flood somewhere in the Middle East.

Miller's problems lay, rather, with evolution, as it was then conceived. In this he was not alone. In that era before Darwin published his more convincing arguments, there was no scientific consensus for evolution, and indeed the scientific establishment opposed *Vestiges* in particular. Miller's arguments were, like those of his contemporaries, partly concerned with the scientific content of *Vestiges*, which was superficial and speculative, and contained some embarrassing errors. That Miller had himself been to the very strata described, and collected and owned many fossils illustrated in the books, gave his words authority; by comparison, the anonymity of *Vestiges*'s author tended to detract from the credibility of the book's arguments.

Vestiges's evolutionism naïvely proposed a simplistic progression of life from simple to complex, so Miller pointedly and tellingly emphasised the presence of large and complicated fossil fishes, such as *Asterolepis*, in very old rocks such as the Old Red Sandstone of Orkney and Caithness. Equally, he could argue that modern fishes such as the sharks had, if anything, apparently greatly degenerated over time, losing the complex bony skeletons and heavy armour observed in early fishes such as *Asterolepis*. Interestingly this notion of fishes' degeneration from the original perfection of Divine Creation, during their time on an imperfect and sinful Earth, harmonised well with Miller's Calvinist world view.

I should note here that when Miller spoke of certain fishes as, for instance, 'sauroid' and close to reptiles, he did not mean that he thought them related in an evolutionary sense. Rather, as was usual in pre-evolutionary times, he followed the old idea of the chain of being in which living things (and indeed all Creation) could be placed in an orderly and unchanging pattern reflecting God's wisdom. This was used to explain the presence of similar structures in related animals, such as the backbone in vertebrates: a phenomenon which evolutionists ascribed to descent from a common ancestor.

Clearly, Miller's objections to evolution were both religious and scientific – which Miller would consider a logical starting position. Evolutionary thought appeared to deny a role for the Divine Creator in the making of living things, and therefore rejected much of natural theology. Miller worried that the spread of evolutionary ideas would lead to the rejection of religion, the collapse of society and eternal damnation for many. There was a further problem. Humans, argued Miller, had souls and were morally responsible for their actions before God their Maker, but animals had neither souls nor moral responsibility. Therefore a gradual transition from animal to human must at some point involve the sudden appearance of the immortal soul – which, to Miller, was preposterous (though many Anglicans, for instance, would have no such problem):

> [If] ... minute vital globules within globules, begot by electricity on dead gelatinous matter ... have at length become the men and women whom we see around us, we must hold either the monstrous belief, that all the vitalities, whether those of monads or of mites, of fishes or of reptiles, of birds or of beasts, are individually and inherently immortal and undying, or that human souls are *not* so.[6]

Miller could see no way out of this dilemma. And if man had no immortal soul, and lacked any life after death, then

> ... what does it really matter to him, for any one moral purpose, whether there be a God or no? If in reality on the same religious level with the dog, wolf and fox, that are by nature *atheists*, – a nature most properly coupled with irresponsibility, – to what one practical purpose should he know or believe in a God whom he, as certainly as they, is never to meet as his Judge? or why should he square his conduct by the requirements of the moral code, farther than a low and convenient expediency may chance to demand?[7]

Miller was running through the kind of arguments which would be repeatedly rehearsed after the debate changed radically in 1859, with the publication of Charles Darwin's *On the Origin of Species*. This book solidly worked through the scientific evidence for biological evolution. Moreover, it provided a credible natural mechanism, 'natural selection', by which animals and plants could evolve without Divine intervention. Miller died in 1856, and therefore never read Darwin's book, which cited Miller's observations on coastal erosion at Cromarty in a famous discussion of the immensity of geological time. It is plainly wrong to imagine Miller as the loser in the crucial evolutionary debate, the one that finally convinced people, simply because the debate never really began in Miller's lifetime. By contrast, back in the 1840s and 50s, it had not been at all clear where the truth lay, especially in the absence of any great scientific debate – let alone a scientific consensus. In that situation, Miller's views were still just about tenable up to 1859; and even after 1859 Chambers's simplistic progressionism was no more 'right' or 'wrong' than Miller's degradatory stasis – both rendered irrelevant by natural selection, which ascribes no significance to the direction of change.

It may well have been sheer bad luck for Miller's posthumous reputation that he died when everything was on the verge of change. We cannot know how Miller would have reacted to Darwin's book – except that he would have studied the issues with his usual care, and no doubt rehearsed his arguments with Lydia over the dinner table. This is because, even with Miller *hors de combat*, the years from 1859 cannot be read as a simple story of Christians versus evolutionists. The vociferous anti-evolutionists included many whom Miller would have despised for their wilful ignorance of science – but also some respected scientists such as Louis Agassiz. The pro-evolutionists were an equally mixed bunch: alongside atheists attracted to Darwinism because of its lack of divine intervention, there were many devout Christians, including Presbyterians who found, no doubt to their surprise, that evolution through natural selection was compatible with Calvinism. As for the middle-of-the-road Christians, many left the issue open, for who were they to criticise God's methods?

Foot-prints, *Testimony* and *Vestiges* remained in print for decades, with new editions appearing in the 1890s; all were by then remarkably out of date, but still attracted a popular readership who found their arguments convincing or reassuring. In any case, there was still real debate and uncertainty even, and perhaps especially, amongst scientists. Darwin's book, and his others that followed, laid the foundation for the modern scientific consensus, from accumulated evidence, that evolution of all life on Earth has taken place, and that this has occurred principally through natural selection. But the nineteenth-century understanding of genetics had badly lagged behind, and this was one reason why even pro-evolutionary scientists needed many decades to assimilate the full breadth of Darwin's thinking, especially on natural selection as the mechanism of evolution. This process was not finished until the middle of the twentieth century. Until then, discussion on evolution was sometimes more reminiscent of Chambersian progressionism, or even Millerian degradation, than natural selection.

Today the similarities between Chambers and Miller seem perhaps more important than the differences. Both insisted, to Chambers's credit – although with reluctance and delay on Miller's part – on debating evolution in the open, when many men of science, most notably Darwin, were frightened to get off the fence and tackle the social and religious implications. Moreover, Chambers's public (if anonymous) fight with Miller and others discharged much lightning which would otherwise have fallen upon Darwin's head, so that the *Origin*'s reception was therefore less ill-considered and more temperate than it might have been.

Indeed, one might well wonder how far Miller's writings contributed to *acceptance* of evolution, for whatever he was discussing, he always encouraged respect for scientific evidence. In yet another of Archibald Geikie's memorably pithy statements, we are told that Miller's unimpeachable Presbyterian credentials had reassured many that geology itself was not, after all, un-Christian: 'His genial ardour and irresistible eloquence swept away the last remnants of the barrier of orthodox prejudice against geology in this country.'[8] Geikie, that master propagandist, was seeking to convince his hearers, at the 1902 centenary celebration of Miller's birth, that there was no longer such a thing as Biblical literalism. But, of course, Miller had not done anything quite so complete, and there would be new battles to come. Meanwhile, the relics of Miller's own campaigning, his fossils reside today in the collections of National Museums Scotland.[9] To view them now, one might think that one was simply looking at scientific objects. Yet, once upon a time, those actual fossils, thanks to their discussion and illustration in Miller's books, symbolised for many people the deepest questions of humanity's origin, purpose and place in the universe.

NOTES TO CHAPTER 15

1. *Footprints*, p.18.
2. *Cruise*, p.175.
3. *Testimony*, p.220.
4. *Footprints*, p.18; 'mechanic' was simply a skilled working man.
5. *Testimony*, p.241; on Miller's scriptural geologists, see O'Connor, *The Earth on show*, chapter 3.
6. *Footprints*, p.13.
7. *Footprints*, p.14.
8. Anon., *Centenary*, p.61.
9. The original '*Asterolepis* of Stromness', found by Miller, is now in Stromness Museum.

HUGH MILLER

Wrapped up warm in his habitual Lowlander countryman's maud.
Original NMS.G.1991.17.1, engraving by Reverend Drummond
of Edinburgh after photograph by James Good Tunny, *c.*1855.

(IMAGE © NATIONAL MUSEUMS SCOTLAND)

CHAPTER 16

A gray maud, buckled shepherd-fashion[1]

IT was not, perhaps, easy to get to know the man behind the editorial façade. When Hugh Miller travelled he went quietly, on foot, or by coach, train or boat. He did not use his contacts with the great and the good of the time to seek accommodation in hunting lodges and great houses; rather, he went incognito, as when he encountered fellow collectors near Gardenstown in Banffshire:

> My reading of the fossils was at once recognised, like the mystic sign of the freemason, as establishing for me a place among the geologic brotherhood; and the stout gentleman producing a spirit-flask and a glass, I pledged him and his companion in a bumper. 'Was I not sure?' he said, addressing his friend: 'I knew by the cut of his jib, notwithstanding his shepherd's plaid, that he was a wanderer of the scientific cast.' We discussed the peculiarities of the deposit ... I showed the younger of the two geologists my mode of breaking open an ichthyolitic nodule, so as to secure the best possible section of the fish. 'Ah,' he said, as he marked a style of handling the hammer which, save for the fifteen years' previous practice of the operative mason, would be perhaps less complete, – 'Ah, you must have broken open a great many.' His own knowledge of the formation and its ichthyolites had been chiefly derived, he added, from a certain little treatise on the 'Old Red Sandstone', rather popular than scientific, which he named. I of course claimed no acquaintance with the work; and the conversation went on.[2]

If one ignores the ephemeral abuse of newspaper controversy, there seems to have survived little criticism of Miller personally. Perhaps Miller simply died too early, so to speak: many accounts were written in the wake of his early death. But another factor was perhaps Miller's apparent wariness and detachment, and his unwillingness to commit himself to taking a full part in wider social intercourse, for better or worse.

To be sure, Miller was no misanthrope. He had friends both geological and non-geological, such as the Liberal peer and philanthropist Lord Kinnaird, the Fife landowner and Free Church supporter David Maitland-Makgill-Crichton, and his minister Guthrie. Miller was, by all accounts, a good friend and a welcome colleague. Nevertheless, it does appear that Miller had little social ambition or appetite for the more formal conventions of society and was, on the contrary, reserved and diffident, even shy, with those whom he did not know. Perhaps he was one of those people who are a little too unsociable for their own good. Although a happy conversationalist with his friends, and a much better listener than the next account implies, he certainly seems to have lacked the arts of small talk to the degree that he found ordinary social events stressful. Lydia recalled:

> When dining out or at an evening party he was usually silent if the conversation were of a gossiping or personal description; but when an opportunity occurred for him to lead conversation in his own way he never failed to embrace it, and the book in his pocket was often brought out to illustrate his ideas. … For any man who had the individuality of genius about him he had an almost exaggerated respect; but the result of passing an evening in the society of commonplace people was always weariness, sometimes to an excessive and painful degree. I never was so sensible of this, however, in Cromarty as afterwards in Edinburgh, where the people were neither intimate friends nor notabilities. These last served to take him out of himself, to rouse and keep him up. It was impossible for him to form any intimate friendship where the intellect did not in a great degree predominate. Thus I should say his closest intimacy in Cromarty was with the widow of an Established minister – a venerable lady resident there after her husband's death … .[3]

Admittedly, refusing ordinary dinner invitations saved him from wearisome distractions, and he had a long list of excuses, such as when he was travelling on Orkney and wanted to save his time for geological wanderings:

... reporting myself a man of irregular habits and bad hours, whose movements could not in the least be depended upon, I had to decline the hospitality which would fain have adopted me as its guest, notwithstanding the badness of the character that, in common honesty, I had to certify as my own.[4]

Miller continued to exercise his old detachment from the formalities of middle- and now upper-class life, even – and perhaps especially – in Edinburgh. His peripheral, even semi-rural, homes outside the town centre seem as consistent with a lack of concern for the social whirl as with a wish to relieve his damaged lungs from the smoky air of the city, or for peace and quiet to help him work and relax. But such a lifestyle does not signal a society-oriented newspaper editor. Miller was certainly not a man whom one can imagine 'working a room' of people like a politician.

Miller had a further reason for his independence. When in 1853 the Duke of Argyll, Cabinet minister, amateur geologist and Free Church supporter, wanted to invite Miller and Guthrie to his ducal seat at Inveraray, he wisely asked Guthrie to mediate. But Miller replied to Guthrie:

I cannot possibly accept of it. It would be easy saying, 'I am not well,' which is at present quite true; and that I am still anxious about Mrs. Miller, which is equally true; but the grand truth in the matter is, that I cannot accept invitations from the great. I feel very grateful for his Grace's kindness. ... But there is a feeling – which, strong when I was young, is now, when I am old, greatly stronger still – that I cannot overcome, and which has ever prevented me from coming in contact with men even far below his Grace's status.

I could easily reason on the point, and have oftener than once done so: – I have said that our nobles have *their* place (and long may they maintain and adorn it), and that I have *mine*, with its own humble responsibilities, and duties; and, further, that men in my position, but vastly my superiors – poor Burns, for instance – have usually lost greatly more than they have gained by their approaches to the great. But I am not to reason the matter, seeing that it exists in my mind mainly as a feeling which I cannot overcome. You will think all this very foolish; but it is fixed, and I really can't help it.[5]

No doubt this had to do with his personal unease and Lydia's illness, but it also points to another aspect of his character, his urge to keep his independence (as already shown in the *Witness* affair), and a refusal to be drawn into the web of obligation spun by the high and mighty of this world. This fits well into a pattern of independence, and even a degree of detachment, from society into which he entered at his own choice of time and place. This starts from his earliest days on the boundary between the middle and working classes, and continues with his experiences as a mason, then as a bank officer – for whom one might imagine it an ideal characteristic – and finally an editor. In his writing, too, there is a sense of that detachment in the way that people are sometimes almost incidental to the landscape.

In the 1840s Miller was the subject of some wonderful photographs by the pioneer Edinburgh photographers David Octavius Hill and Robert Adamson, who used the calotype technique. Miller was fascinated by this new technology and marked the occasion by writing an early but remarkably perceptive critique. Their most famous picture of Miller shows him in shirtsleeves with mell and chisel in hand. But Miller is not in true mason's working clothes; he has simply thrown off his jacket. The photograph is obviously posed by Hill and Adamson in their usual way, as they sought how best to exploit the new technique. Despite – or because of – this artifice, the image is tremendous. Yet the Miller most people saw is better revealed in some of their other photographs which show him in his suit of tweed or some such material, with a maud across his chest, as described in the quote by Geikie at the beginning of this book. In this dress Miller stood out on the streets of Edinburgh. But he was not walking the streets of Edinburgh as a Highlander, or in a hybrid Highland–Lowland fancy dress. Far from it! He was wearing the dress of a Lowlander countryman, as would at once be apparent on market day when the farmers and shepherds were in Edinburgh. The 'maud' which gave warmth, and helped keep rain off, was not the coloured tartan plaid of a Gaelic Highlander, but the grey woollen wrap of a Lowlander shepherd or countryman (more precisely, a pattern known as the Shepherd's Plaid or Border Plaid, of small black and white checks). No wonder Miller was often mistaken for a Lowlander shepherd, especially in the Highlands – which was ironic for such a critic of the Clearances, which were in large part to make room for the Lowland breeds of sheep and their farmers.

Miller's suits seem to have matched their owner's social mutability, befitting the occasion: the good suit of a farmer, perhaps, for city life; but rougher gear, maybe shepherd's tweeds, for a fossil-hunting ramble:

> ... in which I find I can work amid the soil of ravines and quarries with not only the best effect, but with even the least possible sacrifice of appearance: the shabbiest of all suits is a good suit spoiled. My hammer-shaft projected from my pocket; a knapsack, with a few changes of linen, slung suspended from my shoulders; a strong cotton umbrella occupied my better hand; and a gray maud, buckled shepherd-fashion aslant the chest, completed my equipment.[6]

Was Miller consciously or otherwise fashioning himself, perhaps as an actor in his own literary landscapes, dressed like something out of a Walter Scott novel? If one equates 'self-fashioning' with pretence, then obviously not: for a start, there was nothing outlandish about his dress. But in reality dress is important, if only by default, in the projected image which everyone inevitably constructs, whether consciously or not. Miller was, of course, sticking with the comfortable, practical, and above all habitual dress of his Cromarty years – good for keeping dry on the walk home, when an interesting quarry or rocky shore was a far more attractive and relaxing prospect than dressing for formal dinner. It would also match well with his indifference to social matters, and his habit of cutting his own furrow, while his shyness points away from dressing up as some kind of act. Still, even following his old habits, Miller would have known that he stood out on the Edinburgh streets. He was asserting that he was a countryman, by expressing at least a lack of concern for some aspects of middle-class urban social mores. Nor was it without precedent. Take those other literary country lads who came to the Athens of the North: Allan Ramsay, James Hogg and Robert Burns all had their own mauds. But it remains hard to convict Miller of pretence. What one saw was what one got. Rather than some concocted eccentricity, Miller wore the dress of the Scottish Lowlander countryman that he was with the 'pride o worth' of Robert Burns's wonderful song to equality, 'A man's a man for a that':

> What though on hamely fare we dine,
> Wear hodden grey, and a' that?
> Gie fools their silks, and knaves their wine
> A man's a man for a that.[7]

In his independence, his mentality rooted in the burghers of Cromarty, and his detached social position, Miller remained all his life the 'Village Observer' of the title of his juvenile 'newspaper'.

NOTES TO CHAPTER 16

1. *Cruise*, p. 238.
2. *Cruise*, p. 241.
3. L. M. F. F. Miller, 'Mrs Hugh Miller's Journal', p. 371; Mrs Marian McKenzie Johnston kindly identified the lady as Mrs Ann Allardyce, widow of the Revd Alexander Allardyce.
4. *Cruise*, p. 433.
5. Guthrie and Guthrie, *Guthrie*, vol. 2, pp. 325–26.
6. *Cruise*, p. 238.
7. *hamely*, homely; *hodden*, coarse homespun cloth; *a' that*, all that; *gie*, give.

CHAPTER 17

These are but small achievements[1]

A bout of ill-health in 1845 rendered Miller too unwell for his annual tramp around Scotland, so he headed south, not to rest at some comfortable spa, but to spend more than two months travelling, as he narrated in *First Impressions of England and its People* (1847):

'I will cross the Border,' I said, 'and get into England. I know the humbler Scotch better than most men – I have at least enjoyed better opportunities for knowing them; but the humbler English I know only from hearsay. I will go and live among them for a few weeks, somewhere in the midland districts. I shall lodge in humble cottages, wear a humble dress, and see what is to be seen by humble men only – society without its mask. I shall explore, too, for myself, the formations wanting in the geologic scale of Scotland – the Silurian, the Chalk, and the Tertiary; and so, should there be future years in store for me, I shall be enabled to resume my survey of our Scottish deposits with a more practised eye than at present, and with more extended knowledge.'[2]

Miller found the railway network, then beginning its rapid growth, useful for rapid movement, although he wondered about its effects:

One soon wearies of the monotony of railway travelling – of hurrying through a country, stage after stage, without incident or advantage; and so I felt quite glad enough, when the train stopped at Wolverhampton, to find myself once more at freedom and a-foot. There will be an end, surely, to all works of travels, when the railway system of the world shall be completed.[3]

Like *Cruise of the Betsey*, *First Impressions* shows Miller's knack for autobiography in his travel writings. And, like *Cruise*, it is one of Miller's happiest books, full of his delight as he explored the long-imagined haunts of Shenstone, Cowper, and other literary heroes of his youthful reading. He also enjoyed probing the different character of the English, and comparing them to the Scots, sometimes to their credit:

> The English are by much a franker people than the Scotch – less curious to know who the stranger may be who addresses them, and more ready to tell what they themselves are, and what they are doing and thinking; and I soon found I could get as much conversation as I wished.[4]

But the English Sabbath was sadly, if benignly, lax:

> Sunday seems greatly less connected with the fourth commandment in the humble English mind than in that of Scotland, and so a less disreputable portion of the population go abroad. There is a considerable difference, too, between masses of men simply ignorant of religion, and masses of men broken loose from it; and the Sabbath-contemning Scotch belong to the latter category. With the humble Englishman trained up to no regular habit of church-going, Sabbath is pudding-day, and clean-shirt day, and a day for lolling on the grass opposite the sun, and, if there be a river or canal hard by, for trying how the gudgeons bite, or, if in the neighbourhood of a railway, for taking a short trip to some country inn, famous for its cakes and ale; but to the humble Scot become English in his Sabbath views, the day is, in most cases, a time of sheer recklessness and dissipation.[5]

There were museums to visit, and quarries too, as at Wren's Nest at Dudley:

> … I set me down in the sunshine in the opening of a deserted quarry, hollowed in the dome-like front of the hill, amid shells and corallines that had been separated from the shaley matrix by the disintegrating influences of the weather. The organisms lay as thickly around me as recent shells and corals on a tropical beach. The labours of Murchison had brought me acquainted with their forms, and with the uncouth names given them in this

late age of the world, so many long *creations* after they had been dead and buried, and locked up in rock; but they were new to me in their actually existing state as fossils; and the buoyant delight with which I squatted among them, glass in hand, to examine and select, made me smile a moment after, when I bethought me that my little boy Bill could have shown scarce greater eagerness, when set down for the first time, in his third summer, amid the shells and pebbles of the sea-shore.[6]

Many of Miller's writings were, of course, partly autobiographical, because of his inclusion of himself. But he now began a fuller account of his life, using as a starting point his ill-fated *Memoir*. Miller wrote this new autobiography as a series of articles in the *Witness* in 1853, republished in book form as *My Schools and Schoolmasters* (1854), which has become one of his most popular books over the years.

What prompted Miller to tackle this new project was a controversy over the Free Church's educational plans. The dominant clerical faction wanted schools on the traditional parish school model, but exclusive to Free Church members. Miller argued that there should, instead, be a proper state system, for it was more important that children got an education at all (and one still broadly Presbyterian Christian) than that many should be excluded by a strictly denominational system. The actual trigger for the new book came when an opponent asserted that Miller had been unjust to the very educational system which had given him his start in life. Naturally, *Schools*' early chapters focussed on the excellence, or rather the lack thereof, of the traditional Cromarty parish school and its private competitors, none of which were well organised or suited for the majority of their pupils, however successful more diligent or compliant children might be.

Here the full irony of Miller's title started to become apparent, as his story diverged from the standard plot of the legendary (though never mythical) lad o'pairts, perhaps a ploughman's son, who worked hard at school and went on to university and a profession in the apotheosis of the Scottish democratic intellect. Miller subverted this model of pious kailyard poverty by showing how he squandered his early chances. But in telling how he recovered from this debacle, he replaced the standard story with an analysis of the true nature and aims of education, whose freshness can still startle today. Education was much more than formal schooling:

> ... it has occurred to me, that by simply laying before the working men of the country the 'Story of my Education', I may succeed in first exciting their curiosity, and next, occasionally at least, in gratifying it also. They will find that by far the best schools I ever attended are schools open to them all – that the best teachers I ever had are (though severe in their discipline) always easy of access ... .[7]

Education happened in all the different 'schools' of life, not just formal educational institutions. Miller carefully showed, with himself as the example, that there were 'schools' to be had in the family, in one's own reading, in the countryside around, in one's own courtship, marriage and family – and especially in work. Although Miller never advocated purely vocational training, he insisted that education should pay regard to the realities of working life:

> ... that best and noblest of all schools, save the Christian one, [is that] in which honest Labour is the teacher – in which the ability of being useful is imparted, and the spirit of independence communicated, and the habit of persevering effort acquired; and which is more moral than the schools in which only philosophy is taught, and greatly more happy than the schools which profess to teach only the art of enjoyment. Noble, upright, self-relying Toil! Who that knows thy solid worth and value would be ashamed of thy hard hands, and thy soiled vestments, and thy obscure tasks – thy humble cottage, and hard couch, and homely fare![8]

And Miller pointedly reminded the 'higher' classes that 'there are instances in which working men have at least as legitimate a claim to their respect as to their pity'.[9]

Schools became a famous example of self-improvement and making the best of one's life. These were, of course, traditional Scottish aspirations. But they comprised an important Scottish contribution to the 'Victorian values' of modern political legend, thereby matching the spirit of the times so well that Miller became one of the heroes of Samuel Smiles's iconic book *Self-help* (1859). *Schools* sold well, especially in cheap editions, right into the twentieth century. Along with shorter biographies of Miller by other authors, it became a common book prize in schools and bodies such as the Band of Hope, a temperance organisation for working-class children.

Miller's autobiography is remarkable. *Schools* does show its origin in Miller's need to make points in a debate, while the earlier *Memoir* is noticeably more raw, and occasionally perhaps more frank. However, Miller took *Schools* far beyond that controversy's demands, for he presented his life story with real literary skill and effect, almost making a novel of his own life. Of course Miller was hardly an average mason, for his journey would take him far from the chisel and mell. Even as an apprentice, he had a telling encounter with a woman locally regarded as insane:

> For some little time she stood beside me without speaking, and then somewhat abruptly asked, – ' What makes *you* work as a mason?' I made some commonplace reply; but it failed to satisfy her. 'All your fellows are real masons,' she said; 'but you are merely in the disguise of a mason … .'[10]

Nevertheless, Miller had had the advantages and opportunities of any young man in a decent artisan family: no more and no less. Nor was he ever so imprudent as to promise worldly success to his readers. His own story amply showed the role of luck in inflicting accident and illness, but also in giving opportunities to be seized, to move on and upwards. On occasion, perhaps, he seems to wear his heart on his sleeve, such as when he described his youthful rejection of strong drink as a potential turning point. But overall, Miller's dry wit seems, if anything, refreshingly astringent for an age when readers wallowed like Carboniferous amphibians in whole swamps of sentiment. And I certainly do not think that the fifty-something Miller was unduly sentimental when he ended *Schools* with a reflection upon his teenage decision to become a mason:

> In looking back upon my youth, I see, methinks, a wild fruit tree, rich in leaf and blossom; and it is mortifying enough to mark how very few of the blossoms have set, and how diminutive and imperfectly formed the fruit is into which even the productive few have been developed. A right use of the opportunities of instruction afforded me in early youth would have made me a scholar ere my twenty-fifth year, and have saved to me at least ten of the best years of life – years which were spent in obscure and humble

occupations. But while my story must serve to show the evils which result from truant carelessness in boyhood, and that what was sport to the young lad may assume the form of serious misfortune to the man, it may also serve to show, that much may be done by after diligence to retrieve an early error of this kind – that life itself is a school, and Nature always a fresh study – and that the man who keeps his eyes and his mind open will always find fitting, though, it may be, hard schoolmasters, to speed him on in his lifelong education.[11]

Yet that final chapter carried a note of irony to the end, for Miller disposed of his departure for Edinburgh and the successful establishment of the *Witness* in the briefest possible manner, and then ignored the entire subsequent period. It would be logical enough for Miller to end the story of his self-improvement with the final achievement of his presumed aim, while, in any case, he had amply filled one volume. Yet the perceptive reader is left wondering whether the note of ambiguity, of suspended judgement, was deliberate. Did Miller perhaps feel it too early to judge the Edinburgh years, so different from his Cromarty life?

NOTES TO CHAPTER 17

1. *Schools*, p. 561.
2. *First Impressions*, p. 2.
3. *First Impressions*, p. 48.
4. *First Impressions*, p. 65.
5. *First Impressions*, p. 41; the 'fourth commandment', in Reformed Christian tradition, is 'Remember the sabbath day, and keep it holy', Exodus 20:8. Here, 'go abroad' means simply to go out.
6. *First Impressions*, p. 58.
7. *Schools*, p. x.
8. *Schools*, p. 154.
9. *Schools*, p. xi.
10. *Schools*, p. 178; Mrs Marian McKenzie Johnston kindly identified her as Belle (Isobel) Mackenzie, sister of the Revd Laughlan Mackenzie, a well-known preacher.
11. *Schools*, pp. 561–62.

CHAPTER 18

A tenderly affectionate parent[1]

WHEN Miller published *Schools* in 1854 he was already in his fifties; time was passing. Some years before, he noted how a favourite pond in a clearing near his old Conan Mains workplace was now, instead of

> ... the open space and the rectangular pond, a gloomy patch of water in the middle of a tangled thicket, that rose some ten or twelve feet over my head. What had been bare heath a quarter of a century before had become a thick wood; and I remembered, that when I had been last there, the open space had just been planted with forest-trees, and that some of the taller plants rose half-way to my knee.[2]

And the friends of his youth had begun to die off, reminding him, when visiting Cromarty, of 'the old familiar faces' in Charles Lamb's poem of bittersweet nostalgia – as no doubt did his daughter Eliza's grave.[3]

In the 1850s Miller would have had the usual worries of a hard-worked newspaper editor-owner with growing children soon to be sent into the world, and who was supporting his mother and other relatives at Cromarty. He might also have had his disappointments. In 1854 his name was put forward for the vacant Professorship of Natural History at the University of Edinburgh. But Miller, without a university degree, and with few formal scientific papers, was competing against a new phenomenon, the first generation of professional scientists. Bayne – or perhaps rather Lydia – asserted that Miller was deeply disappointed when he lost out to Edward Forbes. But there was no shame in losing to Forbes, an outstanding candidate, and it is not even clear whether Miller himself was disappointed, or had simply been 'volunteered'. In any case, the job was no sinecure, and was widely regarded as killing the ailing Forbes within a year. And

THE MILLER CHILDREN IN 1860

From left to right: Hugh (born 1850), William (1842), Bessie (1845) and Harriet (1839).

(ORIGINAL PHOTOGRAPH IN INVERNESS MUSEUM & ART GALLERY,
HIGH LIFE HIGHLAND [INVMG.1992.224])

in 1855 Miller refused an offer, through Lord Breadalbane, of a well-paid job as a tax administrator, the Civil Service post of Distributor of Stamps and Collector of the Property Tax for Perthshire; there was evidently concern to find him a reasonably paid and less stressful job for his health. His grounds were that he would find it too hard at his age to learn a new job, again claiming his usual lack of confidence in entering a new field. Perhaps, also, he wanted to avoid becoming indebted to the lairds, especially if he was in two minds about becoming a collector of 'taxes on knowledge' (although by the end of 1855 only the paper tax remained).

Still, for someone who had once faced a choice between emigration and poverty, Miller had much to his credit. He had his independence and the regard of his compatriots, and although he was not wealthy he was comfortably off. His worldly success had been almost all his own work, with the support of Lydia (who also had her own literary earnings).[4] Miller and Fairly became clear owners of the *Witness* when Miller discharged his debt to the original subscribers, with a punctilious insistence on full repayment (rather than the 'correct' valuation of the now depreciated printing plant). In February 1854 some of the former proprietors therefore presented him with an engraved salver, 'impressed with the honourable independence of his conduct in regard to that fund and appreciating very highly his eminent services' – perhaps a reference to the affair with Candlish.[5]

Miller had also published several books, with raw materials for still more in the back files of the *Witness*. His literary and geological interests were well established, and he was as much of a public figure as, probably, he ever wanted. He gave geological lectures, often to help raise money for good causes. Prestigious venues included the Philosophical Institution of Edinburgh (some lectures at least stemming from informal geological talks to a group of ladies at his house, whom he also took out into the field),[6] and the Young Men's Christian Association at Exeter Hall in London. But he did not disdain more local audiences and bodies such as the Leith Working Men's Educational Institute,[7] and fundraising events for good causes such as local schools.[8] He also led a geological excursion for 'intelligent mechanics' under the auspices of the Saturday Half-Holiday Association,[9] and supported organisations such as the Scottish Association for Suppressing Drunkenness[10] and the Scottish Young Men's Society.[11]

In June 1848 the Millers had moved to Jock's Lodge, east of Edinburgh, renting (apparently) a house at 2 Stuart Street just off the main road to London in a still largely open area. Here arrived, in 1850, Hugh and Lydia's fourth and last child, Hugh, as Hugh *père* related in a letter to his mother-in-law:

> The doctor has just been with us, and he is well pleased with the appearance of both mother and child. Baby, in his introduction into the world, had a sore struggle for life, and in pugilists' phrase, but with a deeper meaning than theirs, was for about five minutes 'deaf to time'. Accidents can scarce be hereditary; but my mother has told me that, when making my début, I refused to breathe for a still longer period. Were all the future known to the little *entrants* such refusals would, I dare say, be more common … .[12]

And when the census enumerator called on 30 March 1851, the whole family was at home, with two servants from Cromarty and a wet-nurse for wee Hugh.

However, by the mid-1850s Miller would have had serious worries. Lydia was intermittently unwell with an arthritic, sometimes immobilising, condition. He himself was proud of his physical prowess – he would happily join in games of boulder-lifting and table-jumping (but apparently not organised sport, perhaps unsurprisingly), and he could once walk thirty miles a day, in the most manly Victorian manner. But now his health was uncertain (as it may have been for some time, if the 1845 diversion to England is any guide). He was out of action for weeks and months at a time, apparently forced by illness to have a stand-in to deliver his lecture in his presence at a major London hall in February 1854.[13] In May 1854 he had been disabled for months by a 'dangerous chest attack', presumably that which forced him to entrust the *Witness* to a temporary editor for six months.[14] At the beginning of 1855 he was 'unfit for even very slight exertion' after a 'severe and dangerous attack of inflammation of the lungs, which for the greater part of the last eight weeks has confined me to my room'.[15] In December 1855 he had been ill for a month and was still unable to attend a society meeting.[16] In June 1856 he suffered two weeks' attack.[17] When he was due to deliver a talk in Portobello on 23 December 1856, he was evidently so unwell that he was absent and a local minister acted for him.[18] And no doubt more episodes remain unknown. It seems most likely that Miller was suffering from the chronic lung deterioration typical of the silicosis (and perhaps the tuberculosis which often accompanies it synergistically) which he apparently contracted as a mason, and possibly compounded by polluted air and other factors such as stress or asthma. On top of this, Miller suffered acute attacks, probably infections which latched onto his damaged lungs.

It was presumably the relief of paying off the *Witness* that enabled Miller, at last, to buy his own house in April 1854. It was located in Portobello, a village a mile or two further out from Edinburgh than Jock's Lodge and, helpfully for a tired editor, served by omnibuses and North British Railway trains even late at night.[19] Portobello was partly a middle-class seaside resort, and partly a working-class village. But its pottery industry must have added to its attractions, for it meant claypits with fossils. Shrub Mount, the new house, was a villa on the seaward side of the High Street, originally a pleasant eighteenth-century seaside cottage in its own grounds, and extended at least twice before Miller bought it and made further changes. Its 'ceilings were low, and its rooms by no means large; but though it stood close on the High Street it had then an air of comfort and retirement which made it an excellent family residence ... [with a] well-stocked garden'.[20] The garden gave plenty of space for the children, and for Miller to have a free-standing 'museum' built: not in the modern sense of a display institution open to the public, but simply somewhere to house his geological collection. Here he could enjoy talking over his specimens with friends and visitors; at Jock's Lodge in 1850, he had already received such notables as Professor Adam Sedgwick of Cambridge, Sir Roderick Murchison, Professor Richard Owen the anatomist, and Sir Philip Egerton the collector of fossil fish, presumably when they were in Edinburgh for the British Association meeting.[21]

The Miller family now attended Portobello Free Church,[22] although Miller retained links with Free St John's (today Free St Columba's) in central Edinburgh, where he served as a deacon and where his friend Guthrie was minister. In Portobello Miller could take his children to explore the joys of the shore as easily as his Uncle Sandy took him at Cromarty. Bayne recorded that he was 'a tenderly affectionate parent, and never, until the last year of his life, when the serenity of his temper was shaken by reiterated and agonizing attacks of inflammation of the lungs, did he display the least severity to any of his children'.[23] Hugh wrote charming letters to them on his travels; here is one to Harriet from Cromarty:

> I have seen most of our friends here, and found them, as usual, very kind; but I have fenced off all the invitations I could, as I prefer my mother's

potatoes and fish, with full liberty to go about with my hammer at all hours, to much better dinners, and the necessity of being punctual to an hour, and then sitting up quite proper in my chair for some four or five hours at a time.[24]

One cannot but sympathise with Lydia and admire her toleration for such an unconventional husband, with his appalling hours of work, for he was the least punctual or reliable husband at dinner time. Her illness forced her to sleep in a ground floor room to avoid negotiating the stairs, while her husband needed the fresher air on a higher floor, in a separate small bedroom off his upstairs study, which also let him work late without disturbing her.

On the evening of 23 December 1856, Miller helped his daughter Harriet with her school homework by reading to the children a selection of Cowper's poetry. He took a bath in his study upstairs, and retired to bed next door. During the night, Hugh Miller rose, went into the study, raised the hem of his fisherman's jersey, put a pistol to his chest, and pulled the trigger.

NOTES TO CHAPTER 18

1. *Life*, vol. 2, p. 420.
2. *Cruise*, pp. 163–64.
3. *Cruise*, p. 298.
4. More precisely, he also inherited the two Cromarty houses of uncertain value in that declining burgh, although the proceeds from the Leith house and the insurance on his father's sloop were effectively lost in a bad loan and on his poetry: *Memoir*, p. 212.
5. Inscription on tray, Inverness Museum and Art Gallery, INVMG.1992.190.002; *Scotsman*, 8 March 1854, p. 3.
6. *Scotsman*, 31 March 1852, p. 3; *Life*, vol. 2, p. 349.
7. *Scotsman*, 1 April 1854, p. 3.
8. For example, at Newington Free Church in south Edinburgh, for the Causewayside school; *Scotsman*, 8 March 1851, p. 1.
9. *Scotsman*, 14 July 1855, p. 3.
10. *Scotsman*, 17 March 1852, p. 3.
11. *Scotsman*, 7 December 1850, p. 3.
12. *Life*, vol. 2, p. 417.
13. *Life*, vol. 2, pp. 419, 440–41, 461; Miller was also unwell in April 1852 but it is unclear how long or serious this bout was: letter to Robert McKenzie, 17 April 1852, NMS Library SAS Box 616–628 MSS 1929-1.
14. LDGSL/838/M/15/4, Miller to Roderick Murchison, 14 May 1855; *Life*, vol. 2, p. 442; NLS

MS.7516 ff.52, 54; Patrick Dove was temporary editor, *ODNB*.
15. LDGSL/838/M/15/5, Miller to Roderick Murchison, 2 January 1856.
16. NLS Chambers Papers, Dep. 341/86, 'Letters of Noted Persons', 10[48], Miller to Robert Chambers, 15 December 1855.
17. J. A. Smith, 'Notes of fossils from the Old Red Sandstone of the south of Scotland', *Proceedings of the Royal Physical Society of Edinburgh* **2**, 36–37 (1859–62), p. 37.
18. Baird, W. *The Free Church congregation of Portobello: including a sketch of the origin and rise of the town and a history of the church before the Disruption* (Edinburgh, 1889), pp. 101–6, and Baird, W. *Annals of Duddingston and Portobello* (Edinburgh, 1898), p. 481; *Scotsman*, 31 December 1856, p. 2.
19. The year 1854 is taken from sasine of purchase, and kindly confirmed by Mr Archie Foley from editorial addresses in the *Witness*.
20. Baird, *Annals*, pp. 477–78.
21. *Footprints*, pp. 55, 73.
22. Baird, *Free Church*, pp. 105–6.
23. *Life*, vol. 2, pp. 420–21.
24. *Life*, vol. 2, pp. 422–23.

HUGH MILLER MONUMENT, 1953

Miller's statue, with the invariable seagull, looks out to the Sutors over the site of his famous Old Red Sandstone fish discovery, at Cromarty. The monument was erected in 1859 on the fossil sea-cliff behind his birthplace cottage.

(RCAHMS/SCRAN)

CHAPTER 19

Dearest Lydia, dear children, farewell[1]

MILLER'S corpse was found the next morning. His death at first seemed accidental. That it was suicide became clear from the jersey, which was intact, and from a final letter:

> Dearest Lydia,
>
> My brain burns. I *must* have *walked*; and a fearful dream rises upon me. I cannot bear the horrible thought. God and Father of the Lord Jesus Christ, have mercy upon me. Dearest Lydia, dear children, farewell. My brain burns as the recollection grows. My dear, dear, wife, farewell.
>
> Hugh Miller[2]

Guthrie hurried round to the house, where he soon realised the implications:

> In justice both to him [Miller] and to religion, it was considered necessary that a post-mortem examination of the body should be made – that if, as was probable, the brain should be found diseased, that might be made known, and thus, along with other circumstances, remove the last lingering suspicion against Miller which the event might have raised, or his enemies been ready to take advantage of. Mrs. Miller, still ignorant of the real nature of the case, was averse to the body being touched, in the belief, on her part, that his death was purely accidental. In order to get her consent, I had to undeceive her by producing that fond but fatal note which he had left on his desk, addressed to her, expressed in terms of his highest confidence in Jesus Christ, but at the same time plainly intimating his intended purpose, probably executed before the ink on that paper was dry. I shall not soon,

indeed I shall never, forget the face that looked up to mine, and the cry of agony with which the news, though communicated on my part with all possible delicacy, was received.[3]

The brief post-mortem report (signed partly by the same doctors who had treated Miller) spoke vaguely of 'diseased appearances found in the brain', and it was duly found that Miller killed himself 'under the impulse of insanity', when the balance of his mind was temporarily disturbed.[4] It is impossible to know what to make of the 'diseased appearances' today, certainly in modern medical terminology, or whether the doctors' observations, if they weren't simply of some post-mortem artefact, were related to the death at all; but at the time that report was crucial. Long-term mental illness would have cast doubt upon Miller's life and work, and wilfully premeditated suicide would have been deeply shameful, and indeed sinful. At any rate, Miller was humanely allowed a Christian burial, in accordance with the faith shown in his final note, and an outstandingly well-attended funeral procession took him to a grave near Chalmers and other Free Kirk luminaries in Grange Cemetery in south Edinburgh.

Much of our knowledge of Miller's suicide, as of his personal life as a whole, stems from his widow Lydia, mainly in the official biography, *The Life and Letters of Hugh Miller*. Such 'Lives and Letters' were, effectively, deferred funeral orations, parts of the public obsequies for notable Victorians, and were necessarily written – and must be read – within those constraints. Even when they were largely truthful, they would often be somewhat sentimentalised, smoothing off their subjects' rougher edges (while, quite separately, other people's reminiscences might well tend to bask in the reflected glory). Miller's *Life and Letters* is far better than some 'Lives', but it has some special problems. Despite her literary abilities, Lydia was evidently too unwell to take on the task, which went to the journalist Peter Bayne who was, for a while, Miller's successor as editor of the *Witness*. Bayne was no mere ghostwriter, but he only knew Miller personally for the last few years of Miller's life. He must therefore have relied upon, and even been supervised by, Lydia for much of the book, whose delayed publication in 1871 can be ascribed to the delays caused by Lydia's illnesses and spells in various hospitals and spas, as well as Bayne's own disrupted career.

The *Life and Letters*, moreover, saw Miller's entire life through the hindsight of his suicide. Lydia herself wrote a memoir, of unknown date, but at least partly written before 1871 as Bayne evidently used some of it. It is now lost except for those excerpts selected and edited by her granddaughter for publication in 1902, in which Lydia identified the ultimate cause of Miller's suicide as the move from a sedate, relaxed life in Cromarty to a far more stressful one in Edinburgh.[5] But the *Life* told a different tale, plunging from the start into an astonishingly unsympathetic account of Miller's mother (who had died in 1863). She had supposedly filled his childhood mind with demonic tales which, decades later, drove him to his death by 'stimulating the action of a diseased brain'.[6] This profoundly upset Miller's Cromarty family.[7]

One cannot, by the nature of such things, know for sure what really was wrong with Miller. Nor can one always know what to make of the various incidents which were adduced, then and later, as evidence of Miller's mental state. For instance, it might be held significant that Miller drew a pistol on a couple of friends who startled him – but this was many years before, and they were pretending, for a joke, to be ruffians commissioned to beat up the editor of the *Witness*. In fact, Miller had long been in the habit of carrying pistols for self-protection when on his own in lonely areas, first when carrying the Bank's money at Cromarty, or simply when on his geological rambles:

> In the autumn of 1842, during the great depression of trade, when the entire country seemed in a state of disorganization, and the law in some of the mining districts failed to protect the lieges, I was engaged in following out a course of geologic exploration in our Lothian Coal Field; and, unwilling to suspend my labours, had got the pistols, to do for myself, if necessary, what the authorities at the time could not do for me.[8]

Bayne put heavy stress on an episode not long before Miller's death when intruders were noticed in the garden of Shrub Mount, and Miller fretted about his museum, until his neighbour and friend, the Liberal philanthropist Lord Kinnaird, gave him a humane – presumably toothless – mantrap to put in its porch. But Bayne was no fossil collector and wouldn't understand the worries over such a huge investment of time and effort. And when Robert Dick, the fossil-collecting baker of Thurso, talked to a visitor long after Miller's death, he recalled Miller sitting on the ground complaining about 'fairies' getting hold of his trousers, before having a good rub. Samuel Smiles, and many since, took this as evidence for Miller's mental state, when it was most likely a case of deadpan Scots humour on Dick's part, or Miller's – or both.[9]

Some of the worries later adduced by the family, such as the museum mantrap, and Miller keeping weapons to hand in case of burglary, may simply have been the more or less unreasonable frets of an overworked middle-aged man, tired, depressed, and in deteriorating health. But the family were right to point to Miller's health, even if only in hindsight, for we know that he was chronically suffering, and intermittently acutely ill, with the kind of respiratory problems which would have profoundly disturbed his sleep. He had recently, and uncharacteristically, consulted doctors for headaches and for nightmares of the unpleasantly realistic kind which left the victim utterly confused as to what had really happened and what had been a dream, and checking the house or feeling his clothes to see if he had truly been out in the rain and mud.

It is simplest to assume that Miller left it too late to break the spiral of decline – which the doctors' recommendations of rest, warm baths and a haircut did little to halt – until, one night, he suffered an acute episode brought on by a particularly bad nightmare, as hinted by his suicide note. Then, perhaps convinced that he was going mad, he ended everything with the weapon which happened conveniently to be to hand (and which, later, also killed an incautious gunsmith asked to check if it was still loaded). The only oddity was his reading of 'The Castaway' to Harriet in his selection of Cowper's poetry. But this seems a chance consequence of her homework allocation, sandwiched between examples of comedy and pathos as part of a demonstration of Cowper's emotional range. 'The Castaway' is, to be sure, a desperately tragic poem which powerfully embodied Cowper's morbid conviction that he was predestined for damnation. So perhaps Cowper's state of suicidal despair could have been on Miller's mind: but that is not the same as actually planning suicide.

Scottish fairies are, as everyone knows, demonic beings far more dangerous than their butterfly cousins of the English Home Counties, yet there is no evidence that childhood folklore had anything to do with the suicide. One might then wonder whether Miller's imagination contributed to the intensity of his final episode: were his nightmares, perhaps, even more vivid and frightening than another man's? I don't find this a very useful question, if only because it is unanswerable. All the same, it was not unreasonable for Lydia to ask some such question. She could not be objective about her husband, or his suicide. A catas-

Above, left

PROFESSOR LOUIS AGASSIZ

Swiss palaeontologist and world expert on fossil fishes, who wrote about Miller's fossils.

COPY PHOTO FROM E. LURIE, *LOUIS AGASSIZ: A LIFE IN SCIENCE* [CHICAGO, 1960]

Above, right

PATRICK DUFF [detail]

David Octavius Hill and Robert Adamson, calotype.

Town clerk of Elgin, and fossil collector. Miller corresponded with Duff in his last Cromarty years.

SCOTTISH NATIONAL PORTRAIT GALLERY

Left, above

BONES FROM THE HEAD OF A COD WHICH MILLER DISSECTED

Miller studied fresh local fish to understand his fossil fishes better, but this did not answer all his questions because the fossil fishes were so different from the living ones. Largest item $c.$ 11 cm long. Hugh Miller Collection, NMS.

IMAGE © NATIONAL MUSEUMS SCOTLAND

Left, below

BONY PLATES OF A *COCCOSTEUS* LABELLED BY MILLER

Miller was trying to work out their original configuration. NMS.G.1859.33.1050.

IMAGE © NATIONAL MUSEUMS SCOTLAND

1. ONE OF NATIONAL MUSEUMS SCOTLAND'S MOST PRECIOUS OBJECTS

A 'type specimen' upon which the species name of the fish *Pterichthys milleri* is founded. Rock matrix *c.* 7 cm across. Now *Pterichthyodes milleri*. NMS.G.1859.33.5.

2. A TYPICAL CROMARTY NODULE

Head and body armour of *Coccosteus cuspidatus* (see previous page) showing the whole nodule. The fossil fishes were preserved in hard limy nodules whose formation was triggered by the decomposition of the original animal. Nevertheless, the fossils were usually squashed by the weight of overlying sediment. Nodule *c.* 25 cm long. NMS.G.1859.33.1050.

3. ANOTHER CROMARTY FIND

Glyptolepis leptopterus, now considered a sarcopterygian (lobe-finned) fish. One of the Old Red Sandstone fishes collected by Miller at Cromarty. Nodule *c.* 25 cm long. NMS.G.1859.33.1315.

4. A FOSSIL FROM ROBERT DICK AT THURSO

A fossil fish's snout collected by Robert Dick and used by Miller in his writings. Dick later fell on hard times and had to sell his collection; the fossils which Miller described are amongst the few recognisable as Dick's today. Upper jaw termination of *Gyroptychius milleri*, Old Red Sandstone, Thurso, Caithness. Bone *c.* 25 mm across. NMS.G.1859.33.25.

5. DECAYED FROND OF A CYCAD

From Eathie. *Pseudoctenis eathiensis*. About 28 cm long. NMS.G.1911.9.1.

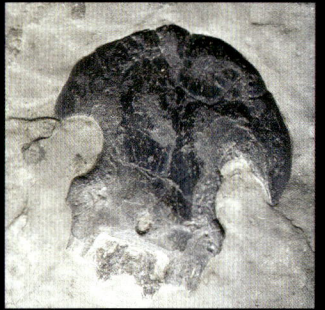

6. EVIDENCE FOR ICE AGE CHANGES OF CLIMATE AND SEA LEVEL

These shellfish came from what was once a cold shallow coastal sea, now land at Fairlie in Ayrshire. Scallop *Chlamys*, with giant barnacle *Balanus porcatus*, 11 cm max. 'Clam' of cold water s--cies '*Astarte elliptica*', now *Arctica islandica*, 7 cm max. Clyde glacial shell beds. NMS.G.1859.33.5005 and 5007.

7. A SERIOUS COLLECTOR'S WORK

Miller collected even incomplete specimens of fossils in his search for information, sometimes in bulk, such as those belemnites. *Pachyteuthis abbreviatus*, Late Jurassic (Oxfordian). Shandwick, north of Cromarty. Up to *c.*12 cm long. NMS.G.1859.33.4028–4042 (various).

8. FOSSIL PLANT FROM THE COAL MEASURES AT MUSSELBURGH

Miller's move to Edinburgh led him to explore the local rocks. *Sphenopteris latifolia*, specimen as seen *c.*15 cm long. NMS.G.1859.33.3232.

6

7

8

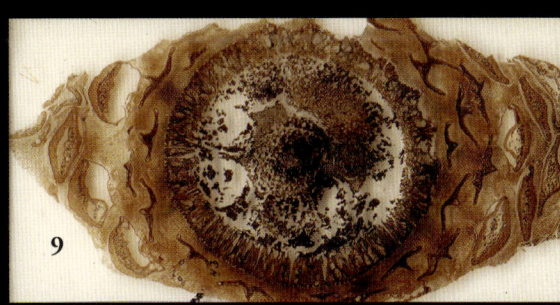

9

9. ONE OF MILLER'S JURASSIC PLANTS

A thin section cut from a 'cone' found at Eathie, and mounted on a microscope slide for research before the First World War. This exquisitely preserved fossil, and other Jurassic plants of Miller's collecting, still intrigue researchers. *Williamsonia scotica*, field of view about 4 cm max. NMS.G.1859.33.4350C.

IMAGES 1–9 © NATIONAL MUSEUMS SCOTLAND

Above

REVD WILLIAM CUNNINGHAM, REVD JAMES BEGG, JOHN HAMILTON AND REVD DR THOMAS GUTHRIE

David Octavius Hill and Robert Adamson

Cunningham, later Principal of the Free Church College, and Guthrie, later minister of Free St John's where Miller worshipped, were friends and allies of Miller, as was Begg, who campaigned for better housing for working people. Hamilton was a lawyer and lay supporter of the Free Church.

SCOTTISH NATIONAL PORTRAIT GALLERY

Right

FREE CHURCH EDINBURGH

David Octavius Hill and Robert Adamson

Calotype of Free St John's (now Free St Columba's) being built on the slopes up to Edinburgh Castle (background), in front of the Established Church's steepled Assembly Hall, September 1844. Miller, through his support for the Free Kirk, helped to generate a multiplicity of ecclesiastical buildings, many now finding new uses as they become redundant.

SPECIAL COLLECTIONS, GLASGOW UNIVERSITY LIBRARY. SCRAN

Above

THE SIGNING OF THE DEED OF DEMISSION
[detail]

David Octavius Hill and
Amelia Paton Hill, 1866

In this part of the famous 'Disruption Painting', Miller is shown reporting at the launch of the Free Church in 1843. Miller is, unusually, wearing a top hat for a formal occasion – although using it as a desk. In any case, he is still in his trademark countryman's maud (wrap) and tweed suit, expressing his pride in his origins and his independence. The mayflower at his feet symbolises the month of May when the Disruption took place, and also mortality: Miller died before the painting was finished.

GEORGE THOMSON;
BY KIND PERMISSION OF THE
FREE CHURCH OF SCOTLAND

A FREE CHURCH SERVICE

A Free Church minister using a boat as a pulpit because of lairds' refusal of sites for new churches. Such open-air services, in the early days of the Free Church, recalled to mind scenes from the Covenanting days and beyond to biblical times. (T. Brown, *Annals of the Disruption with extracts from the narratives of Ministers who left the Scottish Establishment in 1843* [rev. edition, Edinburgh, 1893]).

IMAGE © NATIONAL MUSEUM SCOTLAND

Above

HILL AND FRIENDS

A group calotype including Hugh Miller (with stick) and the photographer David Octavius Hill (sitting), taken 1844–46. Dr George Bell, a close friend of Hill and a founder of the Ragged School movement to educate poor children, is on the right, beyond an unidentified older man, and the landscape painter Macneil MacLeay is wearing what seems to be Miller's maud.

SPECIAL COLLECTIONS,
GLASGOW UNIVERSITY LIBRARY. SCRAN

Left, above

ROBERT CHAMBERS, 1802–1871,
AUTHOR AND PUBLISHER [detail]

David Octavius Hill and Robert Adamson, calotype

Anonymous author of the pro-evolutionary book *Vestiges of the Natural History of Creation*, which Miller criticised.

SCOTTISH NATIONAL PORTRAIT GALLERY

Left, below

OSTEOLEPIS MACROLEPIDOTUS

God's handiwork, according to Miller, was shown in the detailed functional construction of such things as the scales of this fossil fish. From Old Red Sandstone, Cromarty, *c*.17 cm long. NMS.G.1859.33.1218.

IMAGE © NATIONAL MUSEUMS SCOTLAND

Above
MR MILLER AND
MR ROBERTSON [detail]

David Octavius Hill and Robert Adamson, calotype, 1840s. 'Mr Robertson' (on the left) may be the sub-editor of the *Witness* of that name.

SPECIAL COLLECTIONS,
GLASGOW UNIVERSITY LIBRARY. SCRAN

Left
HEAD ARMOUR OF *ASTEROLEPIS*

Head armour of *Asterolepis*, an early but complex fish. Miller used such fossils to challenge *Vestiges*'s naïve view of evolution from simple to complex animals – a telling criticism, but only at that pre-Darwinian time. Head armour (internal view) of *Homostius milleri*, Old Red Sandstone, Thurso, Caithness. Specimen *c*.36 cm long.
NMS.G.1859.33.53.

IMAGE © NATIONAL MUSEUMS SCOTLAND

Above

HUGH MILLER

Oil painting by William Bonnar, of unknown date, but apparently around 1836, from the 'aet. 34' [aged 34] caption to the frontispiece print of Bayne's *Life and Letters*, volume 1, which is based on this painting or a version of it. Another print (by Bonnar and his son Thomas) is dated 1846. NO. INVMG.1992.190.

INVERNESS MUSEUM & ART GALLERY,
HIGH LIFE HIGHLAND

Left, above

THE MILLER FAMILY GRAVE

Hugh and Lydia Miller are buried in Grange Cemetery, south Edinburgh, near Thomas Chalmers and other Free Church worthies.

DR LYALL ANDERSON

Left, below

FAÇADE OF THE HUGH MILLER INSTITUTE

One of the many memorials to Miller, the Cromarty burgh library partly funded by Andrew Carnegie.

DR M. A. TAYLOR

trophe had befallen the family out of the blue. It must have left them shocked, bereaved and guilty, if only in the sense of wondering what they could have done to prevent it. Lydia would, understandably, be searching for any explanation, any rationalisation. But it was grossly unreasonable for Lydia to answer that question about Miller's imagination by blaming his mother for filling his head with fairy stories; and it does not say much for Bayne that he let it pass (especially as he took care to stress that it was Lydia's opinion).

Much weight has been placed on the fact that one of his doctors referred, again after the event, to Miller complaining of feeling, after an atrocious night's sleep, 'as if ... ridden by a witch for fifty miles'.[10] But this seems as unremarkable as an equivalent dream of persecution by a little grey alien would be today. A terrible nightmare will naturally draw upon the dreamer's own stock of horrors, for all that the resulting figurative manifestation remains completely secondary to the original cause of the nightmare. In any case, the same doctor spoiled a literal interpretation by going on to report that Miller had no clear recollection of the content of his nightmare. So Miller might simply have been describing his exhaustion in his usual expressive language.[11]

Whatever its reasons, Miller's death left the family deeply distressed and without their main breadwinner, although Lydia did have some income from her books, and Civil List Pensions were soon awarded to Lydia and to his mother in recognition of his work.[12] It would not have helped that he died intestate. As far as is known, his 'heritable' property, as Scots law calls real estate, comprised Shrub Mount (sold for £875 in 1864) and the two Cromarty houses which could not have been worth much in that run-down burgh. The remainder (the 'moveables') was valued at £7788 9s 10d, of which only a quarter was hard cash, the rest being more or less notional valuations of his fossil collection, of the various copyrights, and of his share in the firm.

The fossils went in 1859 to the Natural History Museum in Edinburgh (precursor of today's National Museums Scotland) for £1025 0s 6d, half paid by public appeal – a better price than Shrub Mount (so that mantrap wasn't so daft after all). The *Witness* was already in somewhat of a decline which has been ascribed to Miller's 'indifferent health, the loosening of the Free Church connection, the adoption of his social opinions and cultural standards by other papers', and his moderation on 'Popery' referred to above.[13] Initially under Bayne's editorship, it continued to decline and was bought up and closed by a competitor in 1864, but it is probably unfair to blame Bayne: the newspaper industry was evolving rapidly with changing tax regimes, increasing mechanisation and new competition from the modern daily newspaper. Miller himself would have had a rough ride.

Lydia never remarried. Despite continuing episodic ill health, physical and mental (perhaps partly caused by Victorian doctors' drug-prescribing habits), she oversaw, as well as *Life and Letters*, the 'preparation of an edition somewhat uniform in size, of the writings of my dear and honoured husband'.[14] *Testimony of the Rocks* was published soon after Miller's death, existing books were reprinted, and new books compiled, sometimes by Lydia, from manuscripts, lecture scripts, and the *Witness*'s files: *Cruise of the Betsey*, *Sketch-book of Popular Geology*, *Headship of Christ*, *Tales and Sketches*, *Edinburgh and its Neighbourhood Essays*, and *Leading Articles*. These books today comprise a selection, edited in places, which is inevitably not strictly representative of the full range of Miller's original writings in the *Witness*. No doubt, if he had lived, Miller himself would have produced more, chosen differently, and rewritten what he did choose. But it is greatly to Lydia's credit that Miller gained a wider readership than he otherwise might have.

Lydia's effort was, of course, to support the family, but it was also in her husband's memory: a 'sacred duty' to give, for instance, his unpublished lectures 'to the world, according to the repeatedly expressed intention of their author'.[15] '[L]oth that any original observation from that mind … should be lost', Lydia creditably fulfilled 'the very difficult and arduous task which God, in His mysterious providence, has allotted me'.[16] She correctly decided to leave Hugh's works much as they were, but with added material, clearly identified as such, to update the older ones such as *The Old Red Sandstone* and *Foot-prints of the Creator*. However, now that Miller was no longer there to react to new discoveries and ideas, notably those of Darwin, his scientific writings dated rapidly, and after a few years Lydia gave up trying to update them.

The children served Church, State and family in the classic manner of the Victorian middle-class. Today their descendants live worldwide, from Cromarty to Australia. Harriet, seventeen when her father died, would have had to bear a heavy burden for her age, compounded by her mother's illnesses. She eventually married John Davidson, who was soon to be ordained as Free Church minister of Langholm in Dumfriesshire before moving to a post at Adelaide, where he became a professor at the university. Harriet had a literary career in her own right, if one haunted by the image of daughters abandoned by brilliant fathers.[17] William became an officer in the Indian Army and retired as a Lieutenant-Colonel. Bessie married a Skyeman, Norman Mackay, who became Free Church minister for Lochinver in 1874, and it was at their manse that Lydia died in 1876. The career of Hugh, the youngest, nicely shows the changes in the scientific world under Victorian reform. Granted a place at the Royal School of Mines in London for his father's sake through Sir Roderick Murchison, he became a field geologist with the Geological Survey, mapping the rocks of northern England and then northern Scotland. He thus came to study the same Cromarty rocks as his father, this time as a professional geologist, a product of the new Government schools and scientific civil service, and, in his spare time, also a pleasant writer on geology. A colleague recalled:

> … it was his great desire to be allowed to map out the district rendered famous by his father, and he afterwards expressed his delight to find that his father had laid down with marvellous accuracy the broad outlines of its geological structure.[18]

NOTES TO CHAPTER 19

1. *Life*, vol. 2, p. 481.
2. Ibid.
3. Guthrie and Guthrie, *Guthrie*, vol. 2, p. 227.
4. *Life*, vol. 2, p. 482.
5. L. M. F. F. Miller, 'Mrs Hugh Miller's Memoir'; for dating, compare, for example, p. 513 with *Life*, vol. 2, p. 120.
6. *Life*, vol. 1, p. 17.
7. Surely a key reason – though not justification – for the tone of Miller's step-nephew H. M. Williamson's memoir, which, apparently in retaliation, portrayed Lydia as neglectful of her wifely duties: see note at end of book.
8. *First Impressions*, p. 245.
9. S. Smiles, *Robert Dick, baker of Thurso, geologist and botanist* (London, 1878), pp. 234–36. Mrs Marian McKenzie Johnston and Dr Ralph O'Connor kindly discussed this; a similar interpretation is made in J. Secord, [review of Shortland (1996)], *British Journal for the History of Science* 31, 249–50 (1998).
10. *Life*, vol. 2, p. 473.
11. Dr David Alston and Dr Ralph O'Connor kindly discussed the problem of folklore and Miller's suicide. Witch-riding is a common motif in Germanic, including Scots, folklore.

12. Lydia, grant 19 June 1857, NLS MS.7516 ff.34–35; Mrs Miller senior, *Scotsman*, 7 November 1857, p. 2.
13. Cowan, *Newspaper in Scotland*, p. 281.
14. *Old Red Sandstone*, 1858 and later editions, p. ix.
15. *Edinburgh*, p. iii.
16. *Sketch-book*, p. xxxvi.
17. For children, also Sutherland and McKenzie Johnston, *Lydia*; Harriet Davidson in *ODNB*; R. B. Walker, 'Davidson, John (1834–1881)', *Australian dictionary of biography*, vol. 4, p. 26 (Melbourne, 1972).
18. J. Horne in Anon., 'The Hugh Miller Celebration', *Transactions of the Inverness Scientific Society and Field Club* **6**, 217–19 (1906), p. 219.

CHAPTER 20

Life itself is a school[1]

HUGH Miller's impact stemmed largely from the opportunity he seized to become editor of the *Witness*. Miller might have had a quieter, perhaps more contented, and very probably longer life, as a Cromarty banker, happy with his family and fossils, and with a modest literary and scientific reputation. By nature, he was a private man. But he had his duty and his opportunity and, for better or worse, he accepted them.

What this impact was upon his time is hard to assess. It is perhaps clearest on geology, measured in the fossils and sites he discovered, the species and genera named after him, and the books he wrote. Miller plainly changed how ordinary people thought about the science. He did not alter the scientists' fundamental theorising – such an impact is only ever vouchsafed to the élite, and fortunate, few – but he did affect them as people: he seemingly captured a generation for geology. Miller had a strikingly analogous impact on Scottish Presbyterianism. He encouraged and mobilised its independent, democratic tradition, to support what became the Free Kirk. But he had relatively little impact on the ecclesiastics themselves, one of whom wrote him out of history.

Miller's suicide, of course, had a huge effect on how he has been perceived. In strictly practical terms this early death cut him off in his literary prime. We have lost what he would otherwise have written, and his work was still developing: one need only compare *Sketch-book* with *The Old Red Sandstone*. As for what he did write, much was never reprinted, while an alarmingly high proportion of his books only made it into print by Lydia's efforts. His death also froze his social and scientific writings at a time of rapid change. This conferred an entirely undeserved reputation as one of the so-called 'scriptural geologists' – the old name for those seen as letting the dogmas of Genesis override their scientific integrity and observation, especially over the age of the Earth – and of

CLOSE-UP OF THE HUGH MILLER MONUMENT, CROMARTY

Miller is carved holding a fossil, his books piled beside him, while the pedestal below his feet is decorated with his eponymous fossil fish *Pterichthys milleri*.

(MICHAEL A. TAYLOR)

whom, ironically enough, Miller was notably critical.

To be fair to Miller, one might well wonder how much his suicide really shortened his working life, given his deteriorating health; but it most certainly was disastrous for his reputation in the way in which it distorted perceptions of his life and work, and not only in the *Life and Letters*. That his actual reasons were as enigmatic as with most suicides has simply left more scope for what sometimes seems the kind of prurient speculation more interested in a man's death than in his life – which, we should remember, was apparently unaffected by mental illness. Indeed, Miller has been perceived as driven to suicide by multiple conflicts. Here he almost seems to be treated as a super-exemplification of the supposed duality of the Scottish mind, with its inherent contradictions – that duality sometimes called the 'Caledonian antisyzygy' by writers such as Gregory Smith and Hugh MacDiarmid: between Highland feyness and Lowland rationality, between science and religion, between progressive and reactionary politics, and so on and so forth. It is almost as if Miller were heir to Robert Wringhim in James Hogg's *Confessions of a Justified Sinner*, and forebear to R. L. Stevenson's *Dr Jekyll and Mr Hyde*. But this is, I submit, nonsense. The evidence is that Miller, as much as anyone can be, was entirely consistent in his world. If anything drove him to death, it was a typically Victorian obsession with work. And for every ill-informed modernist who dismissed him as a loser in the evolution wars, there was seemingly a religious reactionary solemnly wagging a head at Miller's demise, judging from Arnold Bennett's novel *Clayhanger*, set in the Nonconformist Potteries of Staffordshire:

> Even Hugh Miller's *The Old Red Sandstone* ... then over thirty years old, was still being looked upon as dangerously original in the Five Towns in 1873. However, the effect of its disturbing geological evidence that the earth could scarcely have been begun and finished in a little under a week, was happily nullified by the suicide of its author; that pistol-shot had been a striking proof of the literal inspiration of the Bible.[2]

What is most difficult to assess is Miller's impact on the wider body politic. He was no politician or general to be measured in factions organised, votes pushed through, campaigns won, and corpses buried. Nor does it help that much of his journalism was, formally, anonymous. But, quite apart from the

Disruption and its associated issues, Miller must have helped public opinion develop on many matters, such as the Highland Clearances. It is, of course, hard to measure the impact of the way in which Miller spoke directly and individually to each of his readers, even without the complication of the infinitely different ways in which each reader made what he or she wanted of Miller. That would be a study in itself. Nevertheless, it seems to me that Miller perfectly captured the spirit of his time, place and society, especially for hard-working Lowlander Scots seeking to make a living and to understand the world while keeping the self-respect, independence, and Presbyterian faith of their forebears. I have already suggested that he was not far off a moral ideal. And if Miller's impact has proved – so far at least – unquantifiable, that does not mean that it did not exist. He surely changed what it meant, and even today still means, to be a Scot, or a geologist, or a Presbyterian Christian, or simply anyone aspiring to the dignity and worth of education. His critic W. M. Mackenzie – not someone to pull punches – said that

> … though not a great journalist in the technical sense, yet, if he has to be judged by his influence upon public opinion, Miller easily takes a place in the front rank. Probably no single man since has so powerfully moved the common mind of Scotland, or dealt with it on more familiar and decisive terms.[3]

Similar opinions crop up in Miller's contemporaries' accounts. The Revd William Hanna, Chalmers's son-in-law, and joint minister with Guthrie at Free St John's, asserted in what was effectively an obituary that 'Dr Chalmers did not err when, self-oblivious, he spake of Mr Miller, as he often did, as the greatest Scotchman alive after Sir Walter Scott's death …'.[4] That this was not solely due to Free Church partisanship can be judged from Mackenzie's comment and from the quotation from Geikie which begins this book, on how the visitor to Edinburgh 'would be told promptly and with no little pride "That is Hugh Miller". No further description or explanation would be deemed necessary …'.

The regard in which Miller was held by many, then and since (and by implication his impact), certainly seems corroborated by the remarkable variety of monuments to him. His own gravestone is a plain slab of pinkish granite – surprising, perhaps, for a geologist who explored the sedimentary rocks of

Scotland. In 1859, soon after his death, there was set up by public subscription a statue of Miller by Alexander Handyside Ritchie, atop a column on the old sea-cliff above Cromarty. Its inscription reads, 'In memory of Hugh Miller, and in commemoration of his genius, and literary and scientific eminence, this monument is erected by his countrymen'. In the same year, as related above, Miller's fossil collection was given to the nation in his memory; today National Museums Scotland holds Hugh Miller's fossils in trust for present and future generations. Usually some fossils are on display, for visitors to see those evidences for past life, while scientists can examine the collections to make new discoveries and to confirm other scientists' work.

But in a way perhaps the most touchingly personal monument – apart from Lydia's editorial labours – is the 'Disruption Painting' of the First General Assembly of the Free Church by David Octavius Hill and his wife Amelia Paton Hill, based on the calotypes by Hill and Adamson. Miller is prominently placed, may-blossom strewn in front to mark a life cut down in its prime. Amelia Hill also carved a delightful statue, now in National Museums Scotland's collection, of Miller discovering *Pterichthys* on the Cromarty beach. Busts by other sculptors are in the Scottish National Portrait Gallery, the façade of the Falconer Museum at Forres, and the pantheon of Scots heroes in the National Wallace Monument at Stirling.

Aptly utilitarian, in view of Miller's opinions on housing, is Hugh Miller Place of 1862. It was the second street to be built in the pioneering self-help development of working-class housing at Stockbridge in Edinburgh, influenced by his friend the housing campaigner Revd James Begg. Just as appropriate, in its way, is the Alaskan glacier named for him by John Muir, Scots-born American and pioneering nature conservationist – whether in tribute to Miller's self-education, or geological writings, or simply his whole corpus, is unclear, but the glacier pleasingly confirms Geikie's remark that Miller's 'books were to be found in the remotest log-hut of the Far West',[5] and surely Muir agreed with Miller's dictum that 'life itself is a school, and Nature always a fresh study'.[6]

Hugh Miller is, moreover, one of the few geologists in the world with a museum devoted to him, even if his fame depends also on his literary work. His sons, William and (mainly) young Hugh of the Geological Survey, took advantage of the latter's being based in Cromarty for his 1880s fieldwork to establish a small museum in the birthplace Cottage in honour of their father, supplementing a selection of Hugh senior's fossils with some of their own, and others.[7] This museum passed to Cromarty Town Council in 1926, and in 1938 to the National Trust for Scotland, which in 2004 renovated Miller House next door as a splendid new museum about Miller's life and work, enabling the Cottage to be more completely restored to its original appearance. A reorganised and more representative fossil display, together with new interpretation, was sponsored by British Petroleum (BP) plc. Indeed, one BP North Sea oilfield is named after Miller, aptly recalling his comments on the economic importance of geology.

Lydia's editorial labours were vital in developing Miller's posthumous reputation. His books soon stopped being current science. Now they became history, as well as the literature they always were, while much of their enduring appeal must, in any case, have come from their author himself, whatever their varied subjects. They enjoyed sufficiently good sales for a uniform edition to be published in the late 1880s and early 1890s. No doubt a boost to sales came with the 1902 centenary of Miller's birth, which was marked at Cromarty by a gala celebration, at which one speaker was, appropriately, Sir Archibald Geikie, by now a grandee of British science. But Miller would have raised an eyebrow at another speaker, Andrew Carnegie, émigré Scot and ruthless American capitalist – but also an opulent philanthropist who lavishly funded universities, scholarships and public libraries in his native country, including a substantial donation towards Cromarty's new library, the Hugh Miller Institute.

There were further commemorations in 1938, at the handover of the Cottage,[8] and in 1952 at the sesquicentenary of Miller's birth, but as the twentieth century wore on, Miller became increasingly neglected. Geologists continued to love the spirit which imbued his writings, as did Free Church people, while Cromarty folk did not forget their local hero. Historians and anthologists mined his work for his acute observation and trenchant comment. Nevertheless, to many eyes, Miller's writings must have seemed very tired: he was seemingly irrelevant, a controversialist in long-forgotten religious disputes, a failed reconciler – or so it seemed – of Genesis and geology, and a loser in the supposed wars of science versus religion. *My Schools and Schoolmasters* and *The Old Red Sandstone* hung on in print for a while. But, by the 1970s, Miller's books were only to be found in second-hand shops. However, from the 1980s onwards, there came a revival in Scottish history and culture, including folklore, which happened to coincide with a reassessment of the wider nineteenth-century debates in science, geology and religion. The new appreciation of the rich and complex stories in which Miller played a part, and thence of Miller himself, was just in time for the bicentenary of 2002, marked by exhibitions and events in Edinburgh and the Black Isle, and an international conference in Cromarty itself. But one shouldn't rate Miller solely on his impact on the readers of the nineteenth century. In the measured cadences of his speech, all the time attractively rooted in his own person and his own society, Hugh Miller offers great riches to the modern reader in the timelessness of his observations, and the universal questions he poses.

NOTES TO CHAPTER 20

1. *Schools*, p. 562.
2. A. Bennet, *Clayhanger* (London, Penguin edition of 1954), p. 126.
3. Mackenzie, *Hugh Miller*, pp. 238–39.
4. [W. Hanna], [obituary of Hugh Miller], *Witness*, 27 December 1856, p. [2].
5. W. K. Leask, *Hugh Miller* (Edinburgh, 1896), p. 149.
6. *Schools*, p. 562.
7. J. G. Goodchild, *A guide to the geological collections in the Hugh Miller Cottage, Cromarty* (Dingwall, 1902); J. Horne, 'Obituary notice of Hugh Miller', *Transactions of the Edinburgh Geological Society* **7**, 132–38 (1899). See also the special Hugh Miller issue of *The Geological Curator*, vol. **10**, issue 7 (2017).
8. T. S. W[estoll], 'Hugh Miller, 1802–1856. Commemoration at Cromarty', *Nature* **142**, 696–97 (1938); E. Bailey, 'Hugh Miller, 1802–56. Commemoration at Cromarty', *Nature* **170**, 790–91 (1952).

SOURCES

It is impossible in the format and space of such a brief informal biography to give full academic references for every point. Specific notes are, however, given for problematical points and for all quotations, but not matters of fact readily confirmed in the listed sources or in obvious databases (*Oxford Dictionary of National Biography*, registry and census data, sasine registers, local directories, etc.) The following list enumerates the main sources for this book; it is not a full bibliography of Hugh Miller and omission of a publication should not be taken as meaning that it is not worth reading.

REPOSITORIES

Elgin Museum
Inverness Museum and Art Gallery [invmg]
Library, Geological Society of London [LDGSL]
National Galleries of Scotland [NGS]
National Library of Scotland [NLS]
National Museums Scotland [NMS]
Scottish National Photography Collection, Scottish National Portrait Gallery [SNPC]
Stromness Museum

SOURCES FOR BOOK AS A WHOLE

The edition of Miller's works used here is the uniform Edinburgh one of 1880–90s.
Alston, D. 'The fallen meteor: Hugh Miller and local tradition', pp. 206–229 in Shortland (1996).
Alston, D. *Ross and Cromarty. A historical guide* (Edinburgh, 1999).
Alston, D. 'Hugh Miller: the Cromarty years. The social background', pp.10–16 in Borley (2002).
Anon. *The centenary of Hugh Miller being an account of the celebration held at Cromarty on 22nd August, 1902* (Glasgow, 1902).
Bayne, P. *The life and letters of Hugh Miller* (London, 1871), 2 volumes.
Borley, L. (ed.) *Celebrating the life and times of Hugh Miller. Scotland in the early 19th century, ethnography and folklore, geology and natural history, church and society* (Cromarty, 2003; www.cromartyartstrust.org.uk).
Borley, L. (ed.) *Hugh Miller in context: geologist and naturalist: writer and folklorist. A collection of papers presented at two conferences 'The Cromarty Years' (2000)/ 'The Edinburgh Years' (2001)* (Cromarty, 2002).
Brown, C. G. *Religion and society in Scotland since 1707* (Edinburgh, 1997).
Cowan, R. M. W. *The newspaper in Scotland. A study of its first expansion 1815–1860* (Glasgow, 1947).
Gostwick, M. (rev. Powers-Jones, A.) *Hugh Miller's Birthplace Cottage & Museum Cromarty* (Edinburgh, 2013).

Guthrie, D. K. and C. J. Guthrie, *Autobiography of Thomas Guthrie, D. D. and memoir by his sons Revd David K. Guthrie and Charles J. Guthrie, M.A.* (London, 1874–75), 2 volumes.

Knell, S. J. and M. A. Taylor, 'Hugh Miller: fossils, landscape and literary geology', *Proceedings of the Geologists' Association* **117**, 85–98 (2006).

Leask, W. K. *Hugh Miller* (Edinburgh, 1896).

Mackenzie, W. M. *Hugh Miller: a critical study* (London, 1905).

McKenzie Johnston, M. A. 'Miller [*née* Fraser], Lydia Mackenzie Falconer', in *ODNB*.

Macleod, D. 'Hugh Miller, the Disruption, and the Free Church of Scotland', pp. 187–205 in Shortland (1996).

Macleod, D. 'Miller and the Disruption', pp. 80–84 in Borley (2002).

Miller, H. (ed. A. Martin Gostwick) *A noble smuggler and other stories* [early Cromarty journalism] (Inverness, 1997).

Miller, H. *Edinburgh and its neighbourhood, geological and historical: with The geology of the Bass Rock* (Edinburgh, 1891).

Miller, H. *Essays, historical and biographical, political and social, literary and scientific* (Edinburgh, 1890).

Miller, H. *First impressions of England and its people* (Edinburgh, 1889).

Miller, H. *Footprints of the Creator or, The* Asterolepis *of Stromness* (Edinburgh, 1890).

Miller, H. 'Gropings of a working man in geology', *Chambers's Edinburgh Journal*, 28 April 1838, 109–10, and 26 May 1838, 137–39 (reprinted in *Hugh Miller's Memoir*, pp. 246–57).

Miller, H. (ed. M. Shortland) *Hugh Miller's Memoir* (Edinburgh, 1995).

Miller, H. *Leading articles on various subjects* (Edinburgh, 1890).

Miller, H. *My schools and schoolmasters or The story of my education* (Edinburgh, 1893).

Miller, H. *Scenes and legends of the North of Scotland; or, The traditional history of Cromarty* (Edinburgh, 1891, from the 2nd edition of 1850).

Miller, H. *Sketch-book of popular geology* (Edinburgh, 1889).

Miller, H. *Tales and sketches* (Edinburgh, 1889).

Miller, H. *The cruise of the Betsey; or, A summer ramble among the fossiliferous deposits of the Hebrides* (Edinburgh, 1889).

Miller, H. *The Headship of Christ, and the rights of the Christian people* (Edinburgh, 1889).

Miller, H. *The Old Red Sandstone, or, New walks in an old field* (7th ed., Edinburgh, 1899 printing).

Miller, H. *The testimony of the rocks; or, Geology in its bearings on the two theologies, natural and revealed* (Edinburgh, 1890).

Miller, L. M. F. F. (ed. L. M. Mackay) 'Mrs Hugh Miller's Journal', *Chambers's Journal* (6) **5**, 305–8, 369–72, 461–64, 513–16 (1902).

O'Connor, R. *The Earth on show: fossils and the poetics of popular science, 1802–1856* (Chicago, 2007).

Oldroyd, D. R. 'The geologist from Cromarty', pp. 76–121 in Shortland (1996).

ODNB: Oxford Dictionary of National Biography (Oxford, 2004; www.oxforddnb.com).

Rainy, R. and J. Mackenzie, *Life of William Cunningham, D.D., Principal and Professor of Theology and Church History, New College, Edinburgh* (London, 1871).

Secord, J. A. 'From Miller to the Millennium', pp. 328–37 in Borley (2003).

Shortland, M. (ed.) *Hugh Miller and the controversies of Victorian science* (Oxford, 1996).

Stevenson, S. *The personal art of David Octavius Hill* (New Haven, 2002).

Sutherland, E. and M. A. McKenzie Johnston, *Lydia, wife of Hugh Miller of Cromarty* (East Linton, 2002).

Taylor, H. M. 'My recollections of Hugh Miller', pp.163–73 in Sutherland and McKenzie Johnston (2002). For another and more complete version, see that under the pseudonym 'One who knew him', in *British Weekly*, (Scotch Edition), 4 October 1902, pp. 501–2.

Taylor, M. A. 'Miller, Hugh (1802–1856)', in *ODNB*.

Taylor, M. A. 'Introduction', in H. Miller, *The cruise of the Betsey, with Rambles of a geologist*, pp. A13–A52 (facsimile 1st edition, Edinburgh, 2022 revised).

JOURNALS, PAPERS AND PERIODICALS

Annual Reports of the British Association for the Advancement of Science
Archives of Natural History
Ayrshire Histor
British Journal for the History of Science
The Geological Curator see p.169
Inverness Courier
Memoirs of the Geological Survey of the United Kingdom
Nature
Proceedings of the Royal Physical Society of Edinburgh
The Scotsman
Transactions of the Edinburgh Geological Society
Transactions of the Inverness Scientific Society and Field Club
The Witness

ADDITIONAL SOURCES FOR SPECIFIC CHAPTERS

CHAPTER 1
Burnett, J. *A history of the cost of living* (Harmondsworth, 1969).
Conn, S. *Hugh Miller, a one-man play* (Callander, 2002).
Harvie, C. *No Gods and Precious Few Heroes* (Edinburgh, 1998).

CHAPTER 4
Gostwick, M. 'A maestro in the making: Hugh Miller's early writing', pp. 37–48 in Borley (2002).
Hill, B. *The remarkable world of Frances Barkley: 1769–1845* (Sidney, British Columbia, 1978), pp.186–87.
S. J. Knell, The Culture of English Geology, 1815–1851 (Aldershot, 2000)

CHAPTER 5
Waterston, C. D. 'The Cromarty years, Hugh Miller as geologist and naturalist', pp. 26–31 in Borley (2002).

CHAPTER 6
Henderson, L. 'The natural and supernatural worlds of Hugh Miller', pp. 89–98 in Borley (2003).
Robertson, J. 'Scenes, legends and storytelling in the making of Hugh Miller', pp.17–25 in Borley (2002).

CHAPTER 8

Andrews, S. M. *The discovery of fossil fishes in Scotland up to 1845 with checklists of Agassiz's figured specimens* (Edinburgh, 1982).

Collie, M. 'Hugh Miller's dealings with contemporary scientists', pp. 227–36 in Borley (2003).

Janvier, P. 'Armoured fish from deep time: from Hugh Miller's insights to current questions of early vertebrate evolution', pp. 177–96 in Borley (2003).

Torrens, H. S. 'Notes on "The amateur" in the development of British geology', *Proceedings of the Geologists' Association* **117**, 1–8 (2006).

Trewin, N. H. 'Hugh Miller's fish; the "winged *Pterichthys*"', pp. 168–76 in Borley (2003).

Waterston, C. D. 'An awakening interest in geology', pp. 85–91 in Borley (2002).

Waterston, C. D. 'The Cromarty years, Hugh Miller as geologist and naturalist', pp. 26–31 in Borley (2002).

CHAPTER 9

Shortland, M. 'Appendix: a bibliography of Hugh Miller', pp. 301–84 in Shortland (1996).

Shortland, M. 'Hugh Miller's contribution to the *Witness*: 1840–56', pp. 287–300 in Shortland (1996).

CHAPTER 10

Brown, T. *Annals of the Disruption* (new edition, Edinburgh, 1893).

Cheape, H. '"A hotbed of bigotry" and "a sea of difficulties": the Free Church in Hugh Miller's Scotland', pp. 306–20 in Borley (2003).

CHAPTER 11

Henry, J. 'Palaeontology and theodicy: religion, politics and the *Asterolepis* of Stromness', pp. 151–70 in Shortland (1996).

CHAPTER 12

Devine, T. M. *Clanship to crofter's war. The social transformation of the Scottish Highlands* (Manchester, 1994).

Dressler, C. *Eigg. The story of an island* (Edinburgh, 1994).

Fenyő, K. *Contempt, sympathy and romance. Lowland perceptions of the Highlands and the Clearances during the Famine Years, 1845 55* (East Linton, 2000).

Fenyő, K. 'Views of the Highlanders and the Clearances in the Scottish press, 1845 1855: *The Witness* in context', pp. 321 27 in Borley (2003).

Hunter, J. *The making of the crofting community* (Edinburgh, 1995).

CHAPTER 13

Anderson, L. I. 'Hugh Miller: introducing palaeobotany to a wider audience', pp. 63–90 in Bowden, A. J., C. V. Burek and R. Wilding (ed.), *The history of palaeobotany* (Geological Society of London Special Publication **241**, Bath, 2005).

Finnegan, D. 'Edinburgh and the reception of early glacial theory', *Edinburgh Geologist* **40**, 3–11 (2003); www.edinburghgeolsoc.org

Finnegan, D. 'The work of ice: glacial theory and scientific culture in early Victorian Edinburgh', *British Journal for the History of Science* **37**, 29–52 (2004).

Geikie, A. *A Long Life's Work* (London, 1924), pp. 24–25.

Hudson, J. D. 'Hugh Miller's geological discoveries and observations on the Isle of Eigg, as recorded in *The cruise of the Betsey* and in the light of modern knowledge', pp. 197–213 in Borley (2003).

Morrison-Low, A. D. and R. H. Nuttall, 'Hugh Miller in an age of microscopy', pp. 214–26 in Borley (2003).

CHAPTER 14

Geikie, A. *Landscape in History and Other Essays* (London, 1905)

O'Connor, R. 'Hugh Miller and geological spectacle', pp. 237–58 in Borley (2003).

Paradis, J. G. 'The natural historian as antiquary of the world: Hugh Miller and the rise of literary natural history', pp. 122–50 in Shortland (1996).

Secord, J. A. 'From Miller to the Millennium', pp. 328–37 in Borley (2003).

Torrens, H. S. 'William Smith (1769–1839) and the search for English raw materials: some parallels with Hugh Miller and with Scotland', pp. 137–55 in Borley (2003).

CHAPTER 15

Bowler, P. J. *Evolution. The history of an idea* (revised edition, Berkeley, 1989).

Brooke, J. H. 'Like minds: the God of Hugh Miller', pp. 171–86 in Shortland (1996).

Chambers, R. *Vestiges of the natural history of Creation and other writings* (originally published 1844, reprinted 1994 with introduction, etc., by J. A. Secord, Chicago, 1994).

Taylor, M. A. 'Man of Vestiges – Robert Chambers 200 years on', *Edinburgh Geologist* **39**, 32–35 (2002); www.edinburghgeolsoc.org

Secord, J. A. *Victorian sensation. The extraordinary publication, reception, and secret authorship of* Vestiges of the natural history of Creation (Chicago, 2000).

CHAPTER 16

Miller, H. 'The Calotype', *History of Photography* **27**, 7–12 (2003, originally published 1843).

Stevenson, S. 'Hugh Miller's friendship with David Octavius Hill and interest in the photographic art of Hill and Adamson', pp. 72–79 in Borley (2002).

CHAPTER 17

Robb, D. '"Stand, and unfold yourself": My schools and schoolmasters', pp. 246–64 in Shortland (1996).

Vincent, D. 'Miller's improvement: a classic tale of self-advancement?', pp. 230–45 in Shortland (1996).

CHAPTER 18

Baird, W. *Annals of Duddingston and Portobello* (Edinburgh, 1898).

Baird, W. *The Free Church congregation of Portobello: including a sketch of the origin and rise of the town and a history of the church before the Disruption* (Edinburgh, 1889).

Calder, J. 'Professions, proles and profits', pp. 65–71 in Borley (2002).

Campbell, I. and J. Holder, 'Hugh Miller's last house and museum: the enigma of Shrub Mount, Portobello', *Architectural Heritage* **16**, 51–71 (2005).

Smith, J. A. 'Notes of fossils from the Old Red Sandstone of the south of Scotland', *Proceedings of the Royal Physical Society of Edinburgh* **2**, 36–37 (1859–62), p. 37.

CHAPTER 19

McKenzie Johnston, M. A. and M. A. Taylor, 'Lydia Miller and the posthumous reputation', pp. 103–111 in Borley (2002).

Smiles, S. *Robert Dick, baker of Thurso, geologist and botanist* (London, 1878), pp. 234–36.

Taylor, M. A. and M. Gostwick, 'Hugh Miller's collection – a memorial to a great geological Scot', *Edinburgh Geologist* **40**, 24–29 (2003); www.edinburghgeolsoc.org

Walker, R. B. 'Davidson, John (1834–1881)', *Australian dictionary of biography*, vol. 4, p. 26 (Melbourne, 1972).

CHAPTER 20

Anon. 'The Hugh Miller Celebration', *Transactions of the Inverness Scientific Society and Field Club* **6**, 217–19 ([1906]).

Bennet, A. *Clayhanger* (London, Penguin edition of 1954)

Goodchild, J. G. *A guide to the geological collections in the Hugh Miller Cottage, Cromarty* (Dingwall, 1902)

Harvie, C. 'Hugh Miller and the Scottish crisis', pp. 34–47 in Borley (2003).

Taylor, M. A. 'Fellow Scots: John Muir and Hugh Miller', *John Muir Trust Journal and News* **32**, 12–17, 68–69 (2002).

A NOTE ON HUGH MILLER WILLIAMSON'S MEMOIR

Some previous writers, such as Colin MacLean who wrote the libretto for the opera *Hugh Miller* performed at the Edinburgh Festival in 1974, have drawn upon an unpublished memoir of Miller's life written apparently around 1880–81 by Hugh Miller Williamson, son of Andrew Williamson, Miller's half-brother (NLS MS.7527). I do not use it for this book, and this requires an explanation. (See also Sutherland and McKenzie Johnston, *Lydia*, pp.152–53.)

First, Williamson was born in 1855, so he cannot have written from direct knowledge of Miller, and must have relied largely on his relatives. Second, the memoir is tainted evidence. Written apparently in retaliation for Lydia's portrayal of Miller's mother in *Life and Letters*, Williamson's decidedly unpleasant document reveals extraordinary rancour towards Lydia, portraying her as an utterly disastrous wife for Miller. Of course, one should not uncritically accept Lydia's picture of her husband and their marriage in *Life and Letters* (whose limitations, as I say elsewhere, partly reflect its origins); but that does not make Williamson any more reliable. Even allowing for the inevitable rose-tint of hindsight, the independent evidence which we have to date does not support his portrayal. For instance, the memoir by Harriet Taylor gives a very different picture (although she did not shy from criticising Lydia on the separate issue of her treatment of Mrs Miller).

My preliminary assessment is that Williamson's approach stems from the kind of family conflict which sometimes develops between people related by marriage, but otherwise incompatible by temperament, class and lifestyle. I cannot say whether this breakdown of relations happened before Miller's death (though the *Life and Letters* would have been fatal), or whether the memoir's tone came from Williamson, or his elders, or both. The possibility of resentment or grievance arises – but of course does not necessarily follow – from the very real differences in social and economic status, and indeed also generation, involved. Hugh employed his much younger (by more than twenty years) half-brother Andrew on the *Witness*, while Lydia was very much a middle-class lady and employer of servants, in contrast to Andrew's wife Maggie who was recorded on their marriage certificate as a servant in the Cromarty manse, as the late Mr Henry McKenzie Johnston kindly pointed out.

As often happens in such cases, the worst possible interpretation seems to have been placed on whatever Lydia did, even if some more likely explanation offered itself. This is certainly not refuted by sampling, for instance, some points raised by Shortland (himself wary of Williamson's stance: *Hugh Miller and the Controversies*, pp. 43, 46, 49). One needn't doubt Williamson's factual assertion that Miller often slept in Andrew's house in the Newington–St Leonards area of Edinburgh. But his interpretation of Miller avoiding married unhappiness at home discounts the commonsense motive of avoiding the hazards of returning to Portobello in the small hours after the paper had gone to press. That Lydia didn't always have a hot meal waiting seems to reflect not wifely neglect, but the unpredictability of her undomesticated husband – who was in any case happy with a toasted fish and ale – and her own illness. And that Miller chose to convalesce with his mother, rather than at home with the family, surely reflects the quiet, the clean air, and the complete break from work at Cromarty, as well as his regular combination of fossils, family and filial duty, all with commendable economy given that other Victorians in his position would have gone to a spa. The document also shows an apparent immaturity and harshness of judgement, perhaps explained by Williamson's youth.

Williamson's account cannot be accepted at face value, and is useable, if at all, only with caution and as a last resort. However, that is not to take sides, still less to reject it arbitrarily, but to make the pragmatic decision that the job of raking Williamson's bile for pearls of truth is to be tackled only after all other evidence is in. This is partly because the process may in the end be unproductive, and partly because of the need to test Williamson's assertions independently. That demands explicit analysis and justification far beyond the scope of this book.

GLOSSARY

This glossary aims to cover specialised or obsolete words and usages not already dealt with in an endnote and likely to give trouble to the modern, especially non-Scottish, reader. If further reference is needed, *Chambers Dictionary* is especially useful with its inclusion of Scots words.

The technical names of fossil groups and geological terms are defined only informally. It is beyond the scope of this book to explain their often complex relationships to modern concepts and meanings – which may differ radically even when the names are the same. However, it may be helpful to see Oldroyd in Shortland (1997) for geology generally; Janvier and Trewin in Borley (1993) for fishes; and Anderson (2006) for fossil plants.

ammonite: extinct spiral-shelled cephalopod mollusc related to octopuses and pearly nautilus
Araucarian: monkey-puzzle tree or one of its relatives
armour of proof: tested, proven armour
Asterolepis: a type of fossil fish. Miller's material is now known to be a mixture of bones and teeth from two different genera, a mistake such as is common in the early days of researching a new group
Athens of the North: Edinburgh, especially in its eighteenth-century Enlightenment
belemnite: extinct cephalopod mollusc related to modern squid; more properly, the internal skeleton, especially its bullet-like end
bothy: in much of Scotland, the shed housing single male farm workers
brose: oatmeal stirred with water, usually hot; lacking the further boiling to make it palatable
burgh: town with a formally incorporated council (Scots: = borough)
calamite: giant horsetail, now extinct
Celacanths: a group of fossil fishes (Miller could not know of the living coelacanth, discovered in 1938)
Chartism: working-class political movement for universal male suffrage
close: enclosed yard or alley (Scots)
close lanthorn: enclosed lantern
conventicle: open-air prayer meeting, especially of Covenanters (q.v.)
Covenanter: supporter of the Scottish National Covenant of 1638 to defend Presbyterianism; more generally, an opponent of the Stuart kings' anti-Presbyterian policies
cycadalian: cycads, broadly speaking, including extinct forms – i.e. not necessarily from the family to which living forms belong
deacon: a Presbyterian parish officer dealing with secular matters such as finance
Dipterians: group of fossil fishes, including *Dipterus* (today considered a lungfish)
dominie: schoolteacher (Scots)
dress (verb): of a mason, to finish off a block of stone ready for use in a building
enclosed: divided off by hedges and fences, as in the division of previously communally used land amongst local landowners
frith: now usually 'firth': long arm of the sea extending into the land, often as a river mouth (Scots)
Grauwacke: also greywacke, an old name for a type of dark fine-grained sedimentary rock; more generally the rock formation forming much of south Scotland. By extension the period of time during which these rocks were laid down; in Miller's time part of the Silurian, roughly equivalent to the modern Ordovician and (more narrowly defined) Silurian
gryphites: fossil oysters

GLOSSARY

ha-ha: ditch revetted with a vertical wall on the garden side, a landscaping device to keep animals out of the garden of a large house without interrupting the view

hornblende: rock-forming silicate mineral, often found as coarse dark crystals in igneous and metamorphic rock

ichthyodorulite: fossil fin-spine of a fish

ichthyolite: fossil fish

journeyman: hired workman (rather than self-employed artisan)

kailyard: kitchen garden – literally, yard for kale (Scots); figuratively, as in 'kailyard literature', with a sentimental and to some extent mythical view of rural life

kirk: church – both individual building and, especially when capitalised, organisation (Scots)

lad o pairts: talented and promising youth, especially from a humble background (Scots: = 'lad of parts')

laird: a principal landowner in a given area (Scots)

Land o the leal: heaven, paradise (Scots)

Last Judgement: in Christian belief, God's moral judgement of each individual human's soul at the end of time

Lias: usually a geological name for those rock strata now considered the Lower Jurassic; by extension, the equivalent chronological period during which the rocks were laid down. Miller, following the experts of his time, referred the Eathie beds to the Lias. Today they are regarded as later Jurassic

list: (in the motto of the *Witness*) chooses

manse: official residence of a parish minister (Scots)

monad: here, primitive single-celled organism

Nonconformist: in England, relating to a Protestant church, for example Methodist, other than the Church of England; more generally, implying a middle-class independence from the social and political Establishment

offices: on a farm, the working buildings

Old Red Sandstone: the geological name for a particular group of rock strata (by no means all red or all sandstone, as the grey limy nodules in the clayey shales of Miller's Cromarty fish bed show); by extension, the equivalent chronological period during which the rocks were laid down, more or less equivalent to today's usage of the Devonian Period

Oolite: the old name, from the term of a type of granular limestone resembling fish roe, for a group of geological strata roughly corresponding to the modern Middle and Upper Jurassic; the equivalent stretch of geological time

operative mechanic: tradesman or artisan working with his hands

plenishing: household goods, furniture

pre-adamite: relating to the period before Adam and Eve were created

predestination: doctrine that whatever happens is unalterably fixed by divine will; in a Calvinist context, often concerning which souls are to be saved and which damned

processes, vertebral: projecting parts of a vertebra, etc.

puddock: frog or toad (Scots)

recent: old name for the current geological era; especially used to denote a modern animal or plant form as opposed to a fossil one

sasine: relating to obtaining legal possession of property, under the feudal law of land tenure prevailing until recently in Scotland

scaur: bare hillside or cliff

shieling: temporary summer camp or hut used in transhumance, the seasonal grazing of animals on summer pastures distant from the main settlement (Scots)

steading: buildings of a farm, including the working buildings

tawse: implement, often a leather strap, with which Scottish teachers formerly encouraged their pupils with sharp blows, usually on the palm (Scots)

transhumance: see **shieling**

PLACES AND WEBSITES TO VISIT

Below are some worthwhile places to visit. Please note that the mention of other sites in this book does not necessarily indicate safe or legal access, or permission to collect geological specimens such as fossils. Museum opening dates, hours and displays do change, so check before you visit.

CROMARTY is unmissable for anyone interested in Miller. The town and the surrounding countryside are well worth exploring; see in particular:

Hugh Miller's Birthplace Cottage and Museum, National Trust for Scotland. The Birthplace Cottage is furnished in the style of Miller's time, and the Museum in Miller House displays a selection of fossils, some collected by Miller himself; documents, pictures, and personal items such as Miller's wrap and stick; and copies of his book and *The Witness*. Refurbishment of the displays is being planned at the time of writing to make them even more accessible to engage, illuminate and inspire the widest range of visitors.
www.nts.org.uk/visit/places/hugh-millers-birthplace

Cromarty Courthouse Museum combines the fascinating story of an eighteenth century courthouse and jail, complete with cells, prisoners and active court cases, with the story of a vibrant north Highland seaport and its hinterland.
www.cromarty-courthouse.org.uk/

East Church maintained by Historic Churches Scotland [Scottish Redundant Churches Trust]. Formerly Church of Scotland, the kirk was founded in the late 16th century but the building dates largely from the 18th century. The north aisle was added 1739 to create a T-plan church with further alterations in 1756 and 1798. The interior dates from the 18th century with several galleries added in the decades afterwards, most notably in 1756. The church was acquired by the SRCT in 1996, and was restored after being a finalist in the BBC *Restoration Village* series. Following this exposure, it was awarded a major grant in 2007 when the Heritage Lottery Fund, Historic Scotland and The Highland Council paid for a £1.3 million restoration. Following its name change, the Historic Churches Scotland's website was in development as this edition went to press.
https://historicchurches.scot/

EDINBURGH

National Museum of Scotland, Chambers Street. Miller supported its original foundation in the 1850s. Amelia Paton Hill's statue of Miller is in the 'Traditions in Sculpture' gallery. Some fossils collected by Miller, or by others from the same sites, can be seen in the 'Beginnings' gallery, and others are on loan to Hugh Miller's Birthplace Cottage and Museum.
www.nms.ac.uk

PLACES AND WEBSITES TO VISIT

City Chambers, High Street
A commemorative plaque marks the site of the former office of *The Witness*. The printing offices were on Horse Wynd, long since redeveloped; the site is on the west side of the lower part of Guthrie Street, off Cowgate.

76–80 Portobello High Street is Shrub Mount, Miller's last house (there is no public access). The only part of Shrub Mount visible from the High Street is the two-storey house, itself heavily modified since Miller's time, set slightly back behind modern shop fronts; the rear wing, the original eighteenth-century cottage, cannot be seen. The tenement block to the south-east is built on the former front garden, up against Shrub Mount's original frontage which was at right angles to the roadway. The plaque is actually fixed above the doorway to the close which runs through the neighbouring tenement along the original front of Shrub Mount.

Grange Cemetery, Beaufort Road, south Edinburgh. Hugh and Lydia Miller, and other members of their family, are buried in the north-west corner of the main graveyard, near Thomas Chalmers and other Free Church luminaries.

FURTHER WEBSITES

www.thefriendsofhughmiller.org.uk/
The Friends of Hugh Miller is a charity which celebrates and promotes his legacy, and encourages the earth sciences, by broadening public knowledge of Miller's life and work, organising events, and supporting Hugh Miller's Birthplace Cottage and Museum in Cromarty. It publishes *Hugh's News*, with information on events and activities, and articles about Miller. Past copies of *Hugh's News* and membership details can be found on the website.

www.scran.ac.uk
Scottish history and culture, including Miller and some geology

www.scottishgeology.com
Key portal for Scottish geology, including history, fossils and advice on collecting

FURTHER READING

Unfortunately Miller's books are mostly out of print; only those currently, or recently, available are listed here, although others may be available second-hand or online. The books by Devine and Smout are listed for the general Scottish background.

Alston, D. *My little town of Cromarty: the history of a northern Scottish town* (Edinburgh, 2006).

Beake, L. *Jamie's adventures in time. Finding Hugh Miller* (Inverness, 2012). A biography for young people.

Borley, L. (ed.) *Hugh Miller in context: geologist and naturalist: writer and folklorist* (Cromarty, 2002). Interesting papers, complementing the 2002 conference (below) well.

Borley, L. (ed.) *Celebrating the life and times of Hugh Miller: Scotland in the early 19th century: Ethnography and folklore, geology and natural history, church and society* (Cromarty, 2003); also availabe at:
www.cromartyartstrust.org.uk/userfiles/file/Celebrating%20Hugh%20Miller%20sm.pdf [Accessed 3 February 2022]. A collection of papers presented at an international conference held in Cromarty from 10–12 October 2002, to celebrate the bicentenary of Hugh Miller. Organised by the Cromarty Arts Trust in collaboration with the Elphinstone Institute of the University of Aberdeen and the Highland Theological College.

Conn, S. *Hugh Miller*. A one-man play (Callander, 2002).

Davidson, H. M. *Sir Gilbert's Children* (with introduction by H. B. McKenzie Johnston, Inverness, 2011, originally published 1884). By Hugh and Lydia Miller's daughter Harriet, based in part on her memories of family life, if somewhat embroidered.

Devine, T. M. *The Scottish nation 1700–2007* (London, 2006).

Gostwick, M., rev. Powers-Jones, A. *Hugh Miller's Birthplace Cottage & Museum Cromarty* (Edinburgh, 2013). Guidebook and account of Miller.

Gostwick, M. 'Degraded Races, Hopelessly Lost', *Hugh's News, Magazine of The Friends of Hugh Miller*, **46** (November 2020), 11–14.

Miller, H. *A noble smuggler and other stories* (M. Gostwick [ed.], Inverness, 1997). Selected early local journalism.

Miller, H. *My schools and schoolmasters; or The story of my education* (with introduction and notes by J. Robertson, Edinburgh, 1993).

Miller, H. *Scenes and legends of the North of Scotland; or, The traditional history of Cromarty* (revised 1850 edition, with introduction and notes by J. Robertson, Edinburgh, 1994).

Miller, H. *The Cruise of the Betsey, with Rambles of a geologist* (Edinburgh [reprinted 2003, 2022 with introduction and notes by M. A. Taylor, Edinburgh]).

Miller, H. 1841 *The Old Red Sandstone; or new walks in an old field* [forthcoming reprint with critical study and notes, M. A. Taylor and R. O'Connor (eds), Edinburgh].

Miller, H. 1857 *The testimony of the rocks, or, Geology in its bearings on the two theologies, natural and revealed.* [reprinted 2001 with introductions by P. Foster and M. A. Taylor, Cambridge.

Miller, L. *Passages in the life of an English heiress, or, Recollections of Disruption times in Scotland* (with introduction by Elizabeth Sutherland, Inverness, 2011, originally published 1847). Lydia Miller's novel seeking to explain the Disruption to the English.

O'Connor, R. *The Earth on show: fossils and the poetics of popular science, 1802–1856* (Chicago, IL, 2007). Important assessment of Miller as a writer.

O'Connor, R. 'Hugh Miller: racist or anti-racist? Part 1: slavery, the Clearances and Frederick Douglass, *Hugh's News, Magazine of The Friends of Hugh Miller,* **48** (May 2021, revised November 2021), 7–22.

O'Connor, R. 'Hugh Miller: racist or anti-racist? Part 2: scientific racism and the scale of civilization', *Hugh's News, Magazine of The Friends of Hugh Miller,* **50** (December 2021), 2–23

Rosie, G. *Hugh Miller: outrage and order. A biography and selected writings* (Edinburgh, 1981). Useful introductory sampler of the range of Miller's polemic, although Miller's texts are edited. Rosie's account inevitably lacks the benefit of more recent research, but was important in emphasising Miller's role as a journalist.

Reid, L. and Panciroli, E. (eds) *Conversations in stone. Celebrating the legacy of Hugh Miller* (Edinburgh, 2019). An anthology of modern writing inspired by Miller, some from the annual Hugh Miller Writing Competition.

Smout, T. C. *A history of the Scottish people 1560–1830* (London, 1969).

Smout, T. C. *A century of the Scottish people 1830–1950* (London, 1986).

Taylor, M. A., O'Connor, R. and Overstreet, L. K. 2021. Dating the publication of Hugh Miller's *Testimony of the Rocks* (1857). *Archives of Natural History* **48.2**, 310–24.

Taylor, M. A. 2021. 'The unusual printing and publishing arrangements of Hugh Miller (1802–1856)'. *Archives of Natural History* **48.2**, 298–309.

Trewin, N. H. *Fossils alive! Or, new walks in an old field.* (Edinburgh, 2008). How modern geologists interpret the past.

Trewin, N. H. *Scottish fossils* (Edinburgh, 2013).

The Geological Curator **10** (7), August 2017 (free on www.geocurator.org). Hugh Miller Special Edition with a range of papers on Miller's geological collections in his private museum and the museums in Edinburgh and Cromarty over the years, and much else, including family history and biography, the Birthplace Cottage and Miller Monument in Cromarty, early images of the Cottage, and Hill and Adamson's calotype portraits of Miller.

MILLER'S MAJOR BOOKS, WITH DATES OF FIRST PUBLICATION

1829	*Poems, written in the leisure hours of a journeyman mason*
1835	*Scenes and legends of the North of Scotland; or, The traditional history of Cromarty*
1839	*Memoir of William Forsyth, Esq. A Scotch merchant of the eighteenth century*
1841	*The Old Red Sandstone, or, New walks in an old field*
1847	*First impressions of England and its people*
1849	*Foot-prints of the Creator or, the* Asterolepis *of Stromness*
1854	*My schools and schoolmasters; or, The story of my education*
1857	*The testimony of the rocks; or, Geology in its bearings on the two theologies, natural and revealed*
*1858	*The cruise of the Betsey, with Rambles of a geologist*
*1859	*Sketch-book of popular geology*
*1861	*The Headship of Christ, and the rights of the Christian people*
*1862	*Essays, historical and biographical, political and social, literary and scientific*
*1863	*Tales and sketches*
*1864	*Edinburgh and its neighbourhood, geological and historical: with The geology of the Bass Rock*
*1870	*Leading articles on various subjects*

[*posthumous compilations of articles, lectures, unpublished manuscripts, etc.]

INDEX

NB: *Italicised entries denote pictures or illustrations. 'AS1' and 'AS2' denote art sections, with page numbers in brackets.*

actualism, geological 106
Adamson, Robert *28, 52, 124,* 153
Agassiz, Prof. Louis 63, 98–99, 118, *AS2(1)*; fossil fish work 63; glacial theory of 98–99; *Poissons fossiles du vieux grès rouge* 63
Allardyce, Ann, and family 7, 122, 126
Anderson brothers of Inverness (*Guide to the Highlands and Islands*) 60
Anglicans (*see* Church of England)
Anglo-Catholics 83
Apocalypse 62, 104
Argyll, Duke of 123
Atholl, Duke of 87
Baird, Principal 35
Band of Hope, The 131
Barkley, John 37, 42
Bass Rock 109–10
Bayne, Peter 6, 20, 71, 80, 83, 133, 137, 142–43, 145–46
Begg, Revd James 153, *AS2(4)*
Bennett, Arnold 151
Betsey (yacht) 78–79, 92; Miller on 78–79
Bible, The 15, 19, 36, 54, 85, 104, 113, 151; Authorised Version 36; biblical literalism 118
Black Isle 6, 14, 22, 155
Bonnar, William (portrait of Hugh Miller), *AS2(4)*
Border Tales (periodical) 49
bothies/bothy life 29–31, 86, 96
Breadalbane, Lord 135
Brewster, David 49
British Association for the Advancement of Science 63, 65–66, 97, 137
Brougham, Lord 55; 'Letter to …' (by Hugh Miller) 55

Buchanan, Revd Robert 80
Buckland, Professor William 66, 110–11
Burns, Robert 8, 14, 123, 125
Caithness 98, 116
Calvinism 36, 79, 116, 118; Calvinist Presbyterianism 49, 53; Scottish 89
Candlish, Revd Robert 79–80, 135
Carboniferous Period 19, 106, 108; rocks 97
Carnegie, Andrew 19, 154, *AS2(8)*
Carruthers, Robert 35, *AS1(4)*
catastrophes and creations, geological sequence of 62, 104, 106, 129
chain of being 116
Chalmers, Dr Thomas 52, 54, 77, 79–80, 81, 89, 115–16, 142, 152, 167
Chambers, Robert *AS2(6)*; *Vestiges of the natural history of Creation* 114–16, 118–19
Chambers's Edinburgh Journal 49, 65
Chartism 84, 85
Church of England 53, 83
Church of Scotland (The Kirk) 21, 32, 53–56, 69, 72, 75–81, 83; churches (*quoad sacra*) 76; education 54, 129; Established Church 21, 77, 81, 83, 113; kirk session 53, 54; social responsibility 54 (*see also* Free Church of Scotland)
Church, State and Government, union between 77–78
Clearances (*see* Highland Clearances)
climatic change 98–99
Clune quarry 99
coal, Jurassic shale mistaken for 40; coal forests 19; Coal Measures 97, 105, 106, 109–10; Coalheugh Well *102,* 105;

coalfield fieldwork 143; mine, visit to Dryden 107–8
Cobbett, William 75, 90
Commercial Bank 56; at Cromarty 46, 47–48, 59, 60, 84, 143, 149; at Linlithgow 46–47
Conan Mains/Conon Bridge 27, 29, 133
Covenanters 34, 78, 109
Cowper 128, 138, 144
crags-and-tails 98
Creation, biblical, and geology 15, 41, 43–44, 62, 65, 104, 113–19; Genesis 115, 149, 154
CROMARTY: 6–7, 13–15, 22, 23, 27, 33, 34–35, 36, 37, 38, 40, 42, 43, 45–48, 49, 54, 55, 57, 59, 60, 61, 65, 66, 69, 73, 78, 80, *82,* 83, 84, 85, 89, 117, 122, 125–26, 129, 132, 136, 137, 143, 147, 153, 154, 155, 162–63, 166, *AS1(1)*
– Courthouse Museum 166
– economy of 13–14, 33, 47, 86
– Firth 22, 43, 108, *AS1(1, 7)*
– fossils, discovery of 43–4 (*see* fossil fishes)
– geological sketch map *82*
– gravestones by Hugh Miller 34, 51
– Hugh Miller Institute 154, *AS2(8)*
– Hugh Miller Monument 14, *140, 150,* 153
– Hugh Miller Museum 166
– Hugh Miller's Cottage (*see* Hugh Miller's Birthplace Cottage)
– Lodge of the Good Templars 15
– Miller House (*see* Miller, Hugh, Family Homes of)
– Old Parish Kirk (East Church) 53, 78, 166, *AS1(4, 8)*
– schools 7, 129
– streets named after Hugh Miller 14

171

– tourism and Hugh Miller 14, 15
– Town Council 38, 154
Cruise of the Betsey, *with Rambles of a Geologist* 79, 98, 128, 146 (*and* Sources)
Cunningham, Revd William 68–69, 72, 90, *AS2 (4)*
Darwin, Charles 63, 116–19, 146 (*see* Evolution)
Davidson, Lydia (*see also* Middleton, Lydia) 6, 8
Davidson, Revd John (Hugh Miller's son-in-law) 7, 8, 146–47
Day of Judgement 62, 104
Degeneration (of early fishes, *etc*) 116–18
Devonian Period 63 (*see* Old Red Sandstone)
Dick Lauder, Sir Thomas 39
Dick, Robert 98, 143–44, *AS2 (2)*
Disruption, The 13, 75–81, 152, 153; 'Disruption Painting' *AS2 (5)* (*The Signing of the Deed of Demission*: *see* Hill, David Octavius *and* Hill, Amelia Paton)
Divine Creator 62, 113–19 (*see also* Creation)
Dryden 107–8
Dudley 128
Duff, Patrick 61, 65, *82*, *AS2 (1)*
Dunbar of Forres, Miss 6, 39
Eathie 41, 42, 43, *AS1 (6)*; Burn *AS1 (5)*; fossils of the 'Lias' 41, 43–4, 62, 98 , *AS1 (4)*, *AS2 (2, 3)*
EDINBURGH: 8, 17, 31–33, 56, 57, 68, 97–102, 107, 123, 124, 125, 129, 132, 135, 143, 153, 155, 162, 166, *AS1 (2, 3)*; map of 5
– Arthur's Seat 69, 71
– Borough Loch 106–7
– Burdiehouse 100, 107
– Castle Street 32
– Grange 69; Grange Cemetery 142, 167, *AS2 (8)*
– Granton 71
– High Street (City Chambers) 68, 166
– Hugh Miller Place 153
– Jock's Lodge 135
– Leith 31, 72
– Marchmont 69

– Meadows 69–70, 106–7
– Newington–St Leonards 163
– Old and New Towns 32
– Parliament Close, great fire of 32
– Portobello 98, 99, 136, 137, 163, 167; Free Church 136
– Sciennes 69
– University of 35, 133
Edinburgh and its neighbourhood 146 (*and* Sources)
Egerton, Sir Philip 137
Eigg, Isle of 92–94, 101
Elgin 49, 61; Elgin and Morayshire Scientific Association 61, 64
emigration 36, 94, 135; Emigration Stone (by Richard Kindersley) 36
England, visiting 127–29, 136
Episcopalians (*see* Anglicans/ Church of England)
Essays 146 (*and* Sources)
Evangelicals 54, 55, 56, 69, 70, 75, 76, 77, 78; 'Claim of Right' 76
Evolution, Theory of 19, 113–19; Darwinian 63, 116–19, 146; pre-Darwinian 43, 113, 114–16, 118 (*see also* Chambers, Robert)
Fairly, Robert 68, 80, 135
Falconer Museum, Forres 153
First impressions of England and its people 127–29 (*and* Sources)
Fleming, Revd Professor John 61, 114
Foot-prints/Footprints of the Creator 115, 118, 146 (*and* Sources)
Forbes, Edward 133
Forsyth, Isaac 61
Forsyth, William 86; *Memoir of* 49 (*and* Sources)
Forth, Firth of 109
FOSSILS 40–44, 97–101, 113–14, 128–29, 137, *AS1 (6)*
– ammonites 41, 43, *AS1 (6)*
– belemnites 41, 44, *AS1 (6)*, *AS2 (3)*
– bivalves 41, 44, 99, *AS2 (3)*
– calamite 19, 107
– footprints 108–9
– plants 41, 98–99, *AS2 (2, 3)*
– type specimens 63, 100–1

(*see also* Eathie; fossils and other fishes; Hugh Miller and fossil collecting; Hugh Miller and geology, etc; Jurassic; Old Red Sandstone; palaeontology)
FOSSIL AND OTHER FISHES 41, 43–44, 59–66, 99–100, 105, 121, *AS2 (1)*
– '*Asterolepis*' of Stromness 116, 119, *AS2 (7)*
– *Coccosteus* 60–61, 62, 63, *AS2 (1, 2)*
– Dipterians 60
Glyptolepis leptopterus AS2 (2)
Gyroptychius milleri AS2 (2)
– ichthyodorulite 44
– ichthyolites 59, 125; bed 105; ichthyolitic nodule 125
– *Osteolepis macrolepidotus AS2 (6)*
– placoderms 63
– *Pterichthys/Pterichthyodes milleri* 60–62, 63, *150*, 157, *AS1 (7)*, *AS2 (2)*
– sauroid 62, 120
– scales 41, 43–44, 60–1, 113–4
Fraser, Lydia (*see* Miller, Lydia Falconer [*née* Fraser])
Fraser, Mrs (Lydia Miller's mother) 37, 45, 136
Free Church Assembly, first 153
Free Church College 18, 78, 114, *AS2 (4)*; museum of natural science 114
Free Church of Scotland 19, 69, 76–81, 83, 122, 123, 146, 149, 152, *AS2 (4)*; educational plans 78, 133; *Scottish Guardian* (newspaper) 80; worship/ services 78, 79, 85, 93–94, *AS2 (5)* (*see also* Evangelical)
Free St John's Church, Edinburgh 137, 152, *AS2 (4)*
Gaelic (language) 22; Gaels (*see* Highlanders)
Gairloch 29, 96
Gardenstown 121
Geikie, Archibald 17–18, 97–98, 103–4, 106, 119, 124, 152–53, 154; *The founders of geology* 103
General Assembly: (1834) 54; (1843) 76–77; (1843, Free Church) 153

172

INDEX

Geological Society of London 62–63
Geological Survey 18, 98, 147, 153
Girvan 100
Glen Tilt 87
Goldsmith 13
Gordon Cumming of Altyre, Lady Eliza 100
gravestones (*see* Hugh Miller as stonemason …)
Guthrie, Revd Dr Thomas 70–1, 75–76, 80–81, 89, 122, 123, 137, 141–42, 152, *AS2* (*4*)
Hanna, Revd William 152
Harvie, Christopher 18
Headship of Christ 146 (and Sources)
Henderson, Hamish 18
Highland Clearances 36, 94–96, 124, 152; Hugh Miller on 94–96, 152
Highlands and Islands/Highlanders 19, 22, 31, 36, 78, 85, 88, 92–93, 94–96, 124, 151
Hill, Amelia Paton 153, *AS1* (*7*), *AS2* (*5*)
Hill and Adamson, calotypes made by *AS2* (*4, 6*) (*see also* Hill, David Octavius *and* Adamson, Robert)
Hill, David Octavius 28, 52, 124, 153, *AS1* (*2*), *AS2* (*5, 6*)
Hogg, James 125, 151
Hugh Miller: a critical study (by W. M. Mackenzie) 15, 89, 152
Hugh Miller's Birthplace Cottage & Museum (National Trust for Scotland, Cromarty) 6, 22, 23, 38, 49, 145, 153–54, 166, *AS1* (*1*)
Hugh Miller Institute (*see* Cromarty)
Hugh Miller monument (*see* Cromarty)
Hugh Miller's Memoir (by Hugh Miller) 35, 42, 129, 131
Hutton, James 87, 103, 113
Huxley, Thomas Henry 63–64, 67
Ice Ages 98–99, *AS2* (*3*)
Inverness 34, 45–46, 60
Inverness Courier 34–5
Johnstone, John 68, 73
Jurassic, fossil plants 98, *AS2* (*3*); shale 40 (*see also* Eathie)

Kinnaird, Lord 122, 143
Kinnoull, Earl of 55
Kirk, The; kirk session (*see* Church of Scotland)
Knox, John 49, 53, 54, 83, 89
Knox, Robert (*The races of man*) 92
lairds/landowners 19, 30, 32, 36, 53, 54, 77, 78, 84, 85, 86, 87, 89, 94–95, 105, 135
Leading articles … 146 (and Sources)
Leeds, Duke of 87
Leith Working Men's Educational Institute 135
Life and letters of Hugh Miller, The 6, 80, 142–3, 146, 151, 162 (and Sources)
Lowlands/Lowlanders 19, 22, 85, 89, 91, 94, 96, 120, 124, 125, 151, 152
Lyell, Charles (*Principles of geology*) 111
MacDiarmid, Hugh 151
Mackay, Lydia Miller (Hugh Miller's granddaughter) 143; edits grandmother Lydia's memoirs 143
Mackay, Revd Norman (Hugh Miller's son-in-law) 7, 147
Mackenzie, Belle (Isobel) 131, 132
Mackenzie, James 68
Mackenzie, W. M. 15, 89, 152
McNeill, Brian 18
Maitland-Makgill-Crichton, David 122
Malcolmson, Dr John 58, 61–62, 63, 101
Mantell, Gideon 110
maud (plaid) (*see* Hugh Miller and dress)
Melville, Andrew 53
memoir of Hugh Miller (by H. M. Williamson) 162–63
Middleton, Lydia (*née* Davidson) (Hugh Miller's granddaughter) 6, 7, 8
Middleton, Thomas (*see* Foreword) 6, 7
Miller, Elizabeth ('Eliza') (Hugh Miller's first daughter) 50; death of 50; headstone/grave of 51, 133, *AS1* (*5*)
Miller, Elizabeth ('Bessie') (Hugh

Miller's third daughter) 7, 8, 70, *134*, 147; Hugh (Bessie's son) 7
Miller family: papers, loss of many 7; grave of *AS2* (*8*)
Miller, Harriet (Hugh Miller's mother) 22, 23, 24–25, 34, 48, 73, 133, 137–38, 143, 145, 161–62
Miller, Harriet (Hugh Miller's second daughter) 6, 7, 8, 69, *134*, 137, 138, 144, 146–47; *Sir Gilbert's Children* 8
Miller House (National Trust for Scotland, Cromarty) 6, 22, 23, 38, 49, 154, 166, *AS1* (*5*)
Miller, Hugh (Hugh Miller's father) 6, 22; death of 22, 38
MILLER, HUGH
– and access to land 32, 87–88
– and alcohol 15, 31, 32, 33, 131, 135, 163
– and autobiography 18, 66, 121, 129–32, 133 (*see also My schools and schoolmasters*)
– and birthplace 6, 22, 38, 153
– and bothy life 29–31, 86, *AS1* (*2*)
– and Calvinist theology 90, 116
– and childhood/upbringing 6, 22–26, 38, 49, 95–96, 145
– and Church/ministry/religion 53, 55, 77, 78, 79–81, 83, 93, 104, 114, 130, 141–42, 146
– and courtship 45–51
– and The Disruption 75–81 (*see also* The Disruption)
– and Divine Creation 113–20
– and dress 17, 87, 91, 121–26, 127; maud (plaid) 17, 91, *120*, 121, 124–25, *AS2* (*5*)
– and education 6, 18, 19, 23, 24–25, 89, 129–32, 135
– and emigration 45, 140
– and evolution 106, 113–19
– and fatherhood 8, 50, 69, 133, 136–38, 144
– and financial matters 35, 38, 39f, 49, 68, 70, 137, 138f, 145–46
– and firearms 138, 143–44
– and folklore/legend/superstition/tradition 23, 34, 42, 45, 48–9, 66, 107, 143–45, 154
– and fossil collecting 6, 32,

173

41–42, 49, 59–66, 97–101, 125, 128–29, 137, 153; his collection 99–101, 114, 145, 149, 153–54
- and fossil fishes 41–42, 59–66, 113–14
- and game sports 86, 87–88, 89, 92, 94–95; Game Laws 87–88
- and geology/geological writing 19, 40–44, 59–66, 97–101, 103–12, 113–19, 121, 135, 143, 147, 149, 151, 153, 154
- and health/illness 27, 33, 49, 53, 69, 123, 127, 131, 133, 136–37, 141–45, 151
  and the immortal soul 115–16
- and independence/job offers 45–46, 56–57, 79–80, 123–25, 135
- and journalism 19, 23, 35–36, 45, 49, 56, 65, 68–73, 75–76, 77, 78, 105, 151–52
- and local history 37, 48–49, 65
- and marriage 49–51, 61, 71–73, 137–38, 162–63, *AS1* (*4*)
- and microscopy 100
- and mineral collecting 23, 40
- and mining/miners 32, 84, 99, *102*, 105–6, 107–8, 143
- and moral responsibility 117
- and party politics 19, 45–46, 84, 152
- and personal conflict/duality 6, 15, 18–20, 151
- and poetry 6, 34–35, 37
- and race/racism 91–94, 96
- and reputation/impact (contemporary/posthumous) 7, 13, 14, 17–20, 101, 121–22, 141–42, 149–55
- and science and culture 103–12, 113–19
- and science and religion 15, 104, 113–19, 154–55 (*see also* Creation)
- and school (*see* Hugh Miller and education)
- and Scottish history 13, 18–20, 45, 89
- and Scottish patriotism 18, 32, 45, 88
- and social commentary 20, 55, 84–90, 91–96, 130, 143
- and society/social class/position 19–20, 33, 38–39, 50,

84, 85, 90–91, 122, 123–24, 125–26, 127, 129–30, 131
- and Sunday observance 88–89, 128
- and trade unions 32–33, 85
- and travel/travel writings 78–79, 121–23, 127–29
- as banker 46–48
- as burgh councillor 38
- as editor/owner of the *Witness* 6, 56–57, 68–73, 75–81, 91, 105, 123, 124, 129, 133, 134, 135, 136, 142, 145–46, 149
- as scientist (*see* Hugh Miller and geology/geological writing)
  as speaker/accent 22, 24
- as stonemason/monumental mason 6, 19, 25–26, 27–33, 34, 38–39, 40–41, 51, 53, 72, 84, 85–86, 96, 121, 124, 131–32, 133, 136–37, *AS1* (*3, 4*); monumental masonry 34, 51, *AS1* (*2, 4, 5*); sundial 34, *AS1* (*4*)
- as storyteller ('Sennachie') 23
- as writer/literary reputation 18, 23, 34–39, 42, 45–46, 48–50, 56, 65, 69, 70–73, 100–1, 103–12, 118, 135, 146, 149–55
- Bicentenary (2002) 15, 155
- Centenary (1902) 119, 154
- death and funeral 8, 18, 121, 138, 141–47, 149, 152, 167, *AS2* (*8*)
- **FAMILY HOMES OF**: 73, 123; Archibald Place, Lauriston 70; Cromarty (*see* Hugh Miller's Birthplace Cottage *and* Miller House); Shrub Mount, Portobello 99, 137–38, 143, 145, 167; Stuart Street, Jock's Lodge 135–36, 137; Sylvan Place, Meadows 69–70
- Football Club 14
- glacier named for 153
- monuments to *140, 150,* 153–54
- North Sea oilfield named after 154
- Opera, *Hugh Miller* (by R. Barrett-Ayres [music] and C. MacLean [libretto]) 15, 162
- portraits/photographs of *16, 28, 120,* 124, *AS1* (*3, 7*), *AS2* (*5, 6, 7, 8*)

- publications 8, 101, 103, 146, 154 (*and* Sources)
- suicide (*see* death and funeral)
Miller, Hugh (Hugh Miller's second son) 7, *134,* 136, 147, 153–54
Miller, Lydia Falconer (*née* Fraser) 7–8, 22, 37–38, 45–51, 59, 60, 61, 69–70, 71–72, 73, 76, 78, 80, 118, 122, 123–24, 133, 135, 136, 137, 141–47, 153, 162–63; and teaching 45, 47, 49–50; as 'Harriet Myrtle' 72; death of 7, 147; illness 136, 138, 142, 146, 147, 163; 'Mrs Hugh Miller's Journal' (memoir edited by L. M. Mackay) 142–43; *Passages in the life of an English heiress …* 72 (*and* Sources); role in *Life and Letters* 20, 76, 142–44, 162; role in posthumous publication of husband's works 146, 149
Miller, Maggie (Hugh Miller's daughter-in-law) 7
Miller, Jean and Catherine (sisters) 23–24; death of 23–24
Miller, William ('Bill') (Hugh Miller's first son) 7, 70, 129, *134,* 147, 153–54
Moderates 54, 78
monumental masonry (*see* Miller, Hugh as stonemason)
Moray Firth 22, 35, 61, 64, 98, *AS1* (*7*)
Muir, John 19, 153
Muir of Ord 27
Munro family near Lairg (Hugh Miller's uncle, aunt and cousins) 25, 95–96
Munro, George (Hugh Miller's cousin) 25, 95–96
Murchison, Sir Roderick 63, 65–66, 98, 128, 137, 147
museums 128; Hugh Miller's 'museum' 137, 145, 153–54 (*see also* Cromarty: Hugh Miller Museum; National Museums Scotland; Natural History Museum)
Musselburgh 71
My schools and schoolmasters 13, 14, 20, 129–32, 133, 154 (*and* Sources)

National Association for the Vindication of Scottish Rights 88
National Museums Scotland 63, 100, 145, 153, 166; Miller collection 100–1, 119, 145
National Trust for Scotland/Hugh Miller's Birthplace Cottage & Museum 6, 14, 153–54, 166, *AS1* (*1*)
National Wallace Monument, Stirling 153
natural history 114; Professorship of 133
Natural History Museum, Edinburgh 145
Natural History Museum, London (then part of British Museum) 100
natural theology 113, 114, 116
newspapers, conservative 56; liberal 56
Niddrie: Niddrie House 31–32, *AS1* (*2*); Miller's lodgings at 32, 85–86, *AS1* (*2*)
Nigg, kirkyard of 34
Non-Intrusionism 75
Old Red Sandstone: articles about 60, 65–66; fishes 60, 61, 62, 63, 66, 98, 99–100, 113–14, 116, *AS1* (*7*), *AS2* (*6*); Orkney and Caithness 116, *AS2* (*2, 7*); plants 98
Old Red Sandstone, The (book) 65, 66, 105, 106, 121, 146, 149, 151, 154 (*and* Sources)
Oldroyd, Professor David 98
On the origin of species (*see* Evolution)
Orkney 116, 122–23
Owen, Professor Richard 137
Pabay, Isle of 98
palaeontology 40, 101, 114 (*see also* fossils)
patronage, ecclesiastical 53, 54, 55, 77, 88
Paul, Robert 56
Philosophical Institution of Edinburgh 135
photography/calotypy, Miller's comments on 128; (*see also* Hill and Adamson)
Poems written in the leisure hours of a journeyman mason 35, 138

Poor Law, Scottish 87
Presbyterianism/Presbyterians 6, 19, 49, 54, 81, 118, 119, 129, 149, 152; Scottish 149
Rainy, Revd Principal Robert 18
Ramsay, Allan 125
Reform Bill (1832) 38, 84
Reformed/Reformers: 53, 54, 89; Churches 81, 83
Relief Church (1761) 54
religious obscurantism 19
revealed theology 113
Ritchie, Alexander Handyside 153
Robertson, Mr *AS2* (*7*)
Rock, The 83
Roman Catholicism 83–84
Rosemarkie, kirkyard of 34
Ross, Robert 46, 47–48
Ross, William 32
Royal Physical Society of Edinburgh 99, 100
Royal Society of London 18
Rùm, Isle of 94–95; Kinloch Glen 99
Sabbath, The 88–89, 128
St Regulus's kirkyard 51
Saturday Half-Holiday Association 135
Scenes and legends of the North of Scotland; or The traditional history of Cromarty 7, 48–49, 61, 65 (*and* Sources)
Scotland, economy of (agrarian/urban) 84–90, 92–94; map of 4; parish education 129; parish/secular social system 53–54, 81, 87
Scotsman (newspaper) 69, 78
Scott, Walter 32, 34, 35, 36–37, 49, 125, 152
Scottish Association for Suppressing Drunkenness 135
Scottish Guardian (newspaper) (*see* Free Church)
Scottish Young Men's Society 135
Scripture, interpretation of 62, 115–16
Secession Church 54
Sedgwick, Professor Adam 137
Shenstone 132
Silurian Period 131
Sketch-book of popular geology 146, 149 (*and* Sources)

Smiles, Samuel (*Self-help*) 18, 130, 143–44
Smith, Gregory 155
Smith, William 103
Stevenson, R. L. 155
Stewart, Revd Alexander 34, 49, 53
Stuart dynasty 18, 34, 78
Stuart, Mary, Queen of Scots 83
Stuart, Prince Charles Edward (Bonnie Prince Charlie) 18
Suilven (and other Wester Ross hills) 105
Sutherland as it was and is ... 95–96
Sutherland, Duke and Duchess of 95
Sutors 51, *AS1* (*7*); South Sutor headland 43
Swanson, John 42–43, 53, 74, 78–79, 92–93
Tales and sketches 146 (*and* Sources)
Taylor, Esther Ross 7
Taylor, Harriet Ross (Robert Ross's daughter) 47–48, 49, 69–70, 162; Isabella (sister) 47–48
Taylor, Walter 7
Telliamed 43, 113
Ten Years' Conflict, The 80–81
Tertiary 127
Testimony of the rocks (book) 115–18, 146 (*and* Sources)
Testimony of the rocks ... (exhibition, 2002) 9
Thomson, George 8
Times, The (newspaper) 68
Toryism 84
Traditional history of Cromarty, The 46
Union, The (1707) 88
United States of America 104
Vestiges of the natural history of Creation (*see* Chambers, Robert)
Victorians/Victorian values 18, 20, 21, 72, 76, 90, 92, 108, 109–11, 114–15, 130–1, 136, 151
Village Observer, The 23, 126
Wester Ross 105
Whig 79, 84
White, Revd Gilbert, of Selborne 65

Williamson, Andrew (Hugh Miller's half-brother) 162–63

Williamson, Andrew (Hugh Miller's stepfather) 24, 25

Williamson, David (mason and Hugh Miller's employer) 27

Williamson, Hugh Miller (Hugh Miller's half-nephew) 162–63

Williamson, Maggie (wife of Hugh Miller's half-brother) 162

Witness (newspaper) 13, 21, 56–57, 68–73, 75, 78, 79–80, 83–84, 97, 104–5, 124, 129, 132, 135, 136, 137, 142, 145–46, 162–63; editorial offices and print shop 68–69, 97, 166

Wood, Marion (Hugh Miller's friend) 72

Wright, Alexander ('Sandy') (Hugh Miller's uncle) 6, 23, 24, 25–26, 57, 137

Wright, James (Hugh Miller's uncle) 6, 23, 24, 25–26, 30, *AS1(4)*

Wright, Janet (Jenny) (Hugh Miller's aunt) 27

Young Men's Christian Association 135